Be a Successful Green Land Developer

Be a Successful Green Land Developer

R. DODGE WOODSON

New York Chicago San Francisco Lisbon London Madrid
Mexico City Milan New Delhi San Juan Seoul
Singapore Sydney Toronto

The McGraw·Hill Companies

Cataloging-in-Publication Data is on file with the Library of Congress.

Copyright © 2009 by The McGraw-Hill Companies, Inc. All rights reserved. Printed in the United States of America. Except as permitted under the United States Copyright Act of 1976, no part of this publication may be reproduced or distributed in any form or by any means, or stored in a data base or retrieval system, without the prior written permission of the publisher.

McGraw-Hill books are available at special quantity discounts to use as premiums and sales promotions, or for use in corporate training programs. To contact a special sales representative, please visit the Contact Us page at www.mhprofessional.com.

1 2 3 4 5 6 7 8 9 0 DOC/DOC 0 1 4 3 2 1 0 9 8

ISBN 978-0-07-159259-8
MHID 0-07-159259-8

Sponsoring Editor
Joy Bramble Oehlkers
Acquisitions Coordinator
Rebecca Behrens
Editorial Supervisor
David Fogarty
Project Manager
Jacquie Wallace
Copy Editor
Wendy Lochner

Proofreader
Leona Woodson
Indexer
Leona Woodson
Production Supervisor
Pamela A. Pelton
Composition
Lone Wolf Enterprises, Ltd.
Art Director, Cover
Jeff Weeks

 The pages within this book were printed on acid-free paper containing 100% postconsumer fiber.

Information contained in this work has been obtained by The McGraw-Hill Companies, Inc. ("McGraw-Hill") from sources believed to be reliable. However, neither McGraw-Hill nor its authors guarantee the accuracy or completeness of any information published herein, and neither McGraw-Hill nor its authors shall be responsible for any errors, omissions, or damages arising out of use of this information. This work is published with the understanding that McGraw-Hill and its authors are supplying information but are not attempting to render engineering or other professional services. If such services are required, the assistance of an appropriate professional should be sought.

*This book is dedicated to my long-time friend, Larry Hager,
who has been involved in my writing career for some 17 years.
I owe a great deal to this icon of the publishing business.
Larry—thanks, man.*

Contents

Introduction — xvii

CHAPTER ONE
Why Should I Consider Green Land Development? — 1

How Much Money? — 2
How Do I Find Partners? — 4
More Control — 5
How Much Risk? — 6
How Can I Compete? — 6
Red Tape — 7
The Transition from Builder to Developer — 8

CHAPTER TWO
Using Green Space to Make Big Bucks — 9

Why Will People Pay More? — 10
Which Types of Features? — 10
How Much Am I Going to Lose? — 11
Get in Early — 11

CHAPTER THREE
Building a Strong Development Team — 13

Ideas	14
Money and Credit	15
Seller Financing	15
Options	16
Seller Participation	16
Organizational Skills	16
Research	17
People Skills	18
Time	18
Choosing Your Development Team	19
The Experts	20
Attorneys	21
Accountants	21
Marketing Consultants	21
Builders	21
Project Planner	22
Architects	22
Civil Engineers	22
Surveyors	23
Finding the Right People	23
Questions to Ask	24
A Rough Plan	24
Lenders	25

CHAPTER FOUR
Making Sales — 27

Independent Sales	28

Brokerage Sales	29
Listing Agreements	30
Being Your Own Salesperson	31
Your Sales Team	31
A Marketing Plan	32

CHAPTER FIVE
Setting Up Your Lenders and Investors — 35

Commercial Banks	37
Mortgage Brokers	38
Private Investors	39
Builders and Contractors	40
Partnerships	41
Creative Terms	41

CHAPTER SIX
Finding Suitable Development Property — 45

Availability of Land	46
Defining Your Needs	47
Setting a Budget	49
Working with Brokers	51
The Do-It-Yourself Approach	52
Appraisers	52
Selecting Viable Projects	53
Hard Sells	54
Competition	54
Average Deals	55
The Mass Market	56

Fine Tuning	57
Going Around in Circles	58

CHAPTER SEVEN
What Makes a Project Viable — 59

Components	60
Small Developments	61
Your Job	61
You Make the Rules	63
Land Use	63
Building Regulations	64
Construction Regulations	65
Temporary Construction	65
Site Needs	66
Project Manager	67

CHAPTER EIGHT
Looking For Land in All the Right Places — 69

Dealing Directly	70
Back Taxes	72
Auctions	72
Foreclosures	73
Lawyers	74
Accountants	74
Benefits	74

CHAPTER NINE
Initial Investigations — 77

Space Preservation	77

Will You Combine Mixed-Use Properties?	78
Urban Infill and Brownfield Redevelopment	78
Preplanned Growth	78
Erosion Factors	79
Irrigation	79
Summing Up	79

CHAPTER TEN

Drainage Factors — 81

Closed Drainage Systems	83
Open Systems	84
Looking at the Land	86
Sectional Designs	86

CHAPTER ELEVEN

Soil Considerations — 89

Types of Soils	90
Bearing Capacity	91
Compaction Equipment	93
Sheepsfoot Rollers	93
Smooth Rollers	93
Rubber Rollers	94
Vibrating Rollers	94
Power Tampers	94
Stability	94
Importance of Soil Assessment	95

CHAPTER TWELVE

Calculating Land Loss from Road Costs and Green Space — 97

Simple Access Roads	98

Straight and to the Point	99
Branches	99
Natural Road Sites	100
Complex Roads	100
Green Space	102

CHAPTER THIRTEEN
Water Requirements — 105

Types of Water Demand	106
Routing Water Mains	108
Fire Hydrants	109
Tapping In	110
Wells	111

CHAPTER FOURTEEN
Flood Zones, Wetlands, and Other Deal-Stoppers — 113

Working with Wetlands	115
Flood Areas	117
Hazardous Waste	118
Other Environmental Concerns	118
Angry Citizens	119

CHAPTER FIFTEEN
Location, Location, Location — 121

Matching Up Your Needs	122
Demographics	124
Traffic Studies	126
Proximity Research	127

Failed Projects — 128
Growth Status — 128

CHAPTER SIXTEEN
Plans and Specifications — 131

Plan Layers — 131
Types of Plans — 133
    The Demolition Plan — 133
    The Grade Plan — 134
    Utility Plans — 134
    The Traffic Plan — 134
    The Construction Plan — 135
Building Plans and Specifications — 135
Changes — 136
Cutting Costs — 137

CHAPTER SEVENTEEN
Working with Contractors — 139

Estimates Are Not Quotes — 140
Locating Contractors — 142
Soliciting Bids — 143
Comparing Bids — 144
Refining Bids — 145

CHAPTER EIGHTEEN
Projecting Profit Potential — 147

Making Sales Projections — 147
Local Trends — 148

Contents

Working with a Team	149
Land Only	150
Land and Improvements	151
Beware of Hype	152

CHAPTER NINETEEN

Zoning Considerations 155

Zoning Maps	155
Cumulative Zoning	157
Floating Zones	158
Transitional Zoning	158
Planned Unit Developments	159
Cluster Housing	159
Exclusionary Zoning	160
Inclusionary Zoning	160
Other Types of Zoning	161
Matching Up Your Needs	122

CHAPTER TWENTY

Closing Deals 163

The Closing Process	164
The Appraisal Process	165
The Title Search	166
Surveys	167
Income Verifications	167
Credit Reports	167
Tracking	168

CHAPTER TWENTY-ONE

Supervising Your Site — 171

Going Solo — 171
What Qualities Should I Look For? — 172
What Will You Expect? — 173
The Decision Is Yours — 174

CHAPTER TWENTY-TWO

Staying on Budget and on Time — 175

Tracking Your Budget — 176
    Routine Reports — 177
    Invoices — 177
    Price Increases — 178
Staying on Schedule — 178
    Supervision — 180
    Organization — 181
    Communication — 182
    Legwork — 182
    Control — 183
    Money — 183
    Backup Plans — 184
    Another Project — 184

CHAPTER TWENTY-THREE

Ideas for Environmentally Friendly Developments — 87

Key Considerations — 188
Partners — 189
Pedestrians — 190

General Design ... 190
Street Solutions ... 190
Grass ... 191
Ride-Sharing Parking Lots ... 191
Mixed-Use Ideas ... 191
Preservation ... 192
Natural Wind Breaks ... 193
On-Site Wastewater Treatment ... 193
Covenants and Restrictions ... 194
Collect Brochures ... 195
Talk with Bankers ... 195
Real-Estate Professionals ... 195
Interview Development Professionals ... 196
Reach Out ... 196
Bottom Line ... 196

CHAPTER TWENTY-FOUR

Going from Green Developing to Green Building   199

20-Acre Mini-Estates ... 199
What Does It Take to Become a Builder? ... 200
Do I Have to Be a Carpenter? ... 201
What Does a Builder Do? ... 202

Appendix 1: Glossary of Green Words and Terms   205

Appendix 2: Time Saving Tips and Tables for Land Developers   227

Index ... 337

Introduction

Sustainable building practices are the wave of the future. Going "green" is good for the environment and can be very good for your bank account. The demand for eco-friendly building practices is high. People are willing to pay more to waste less. This is a key to future success for land developers.

Whether you are a carpenter or a builder who wants to venture into becoming a green land developer, you should give serious consideration to sustainable development and construction practices.

Going green is not as simple as using a few green tactics in your development of house lots and commercial spaces. You should integrate various sustainable options to provide a full green package. This is the best way to earn your reputation as a green developer.

Before you take out new ads that label you as a green developer, you need to understand what responsibilities you are taking on. To be a real green developer, you are going to need to educate yourself from start to finish, and this book is the starting point.

With over 30 years of experience as a developer, designated real estate broker, builder, remodeler, and master plumber, I know what I'm talking about. This is your chance to gain invaluable experience for a fraction of the cost of what a one-hour consultation would cost. If you are looking for a brighter future in the green development business, this book is the book that you need.

Take a look at the table of contents. Flip through the pages and notice the tip boxes. I've done everything possible to make this book a pleasant trip rather than a boring learning experience. I trust that you will find this an effective view of what it takes to become a successful green land developer without the suffering often associated with learning. This is your fast track to getting in on the green boom.

Be a Successful Green Land Developer

CHAPTER ONE

Why Should I Consider Green Land Development?

Why would anyone consider green land developing? Money can be a major motivator, and there can be a lot of money in developing raw land. The profit potential of land development is incredible. When you add the twist of developing with sustainability as a cornerstone, profit potential grows. There is substantial risk attached to some development deals, but there is usually risk when there is the potential of rich rewards. Homebuilders need land to build on. Most builders buy their building lots from developers, but they can increase their profits considerably by developing their own building lots. How much extra money can builders make when they develop their own lots? It depends on many factors, but some builders have cut their land cost in half. I know that this is true, because I am one of those builders.

Green builders and green developers need each other. The total concept of environmentally friendly developing needs both builders and developers to be on the same track. Green builders are not going to do nearly as well building in standard developments as they are when they work with green developers. In a similar vein, green developers will be gambling on the sustainability factor when they go green. If the builders who buy the developed lots are not willing to pay for the added benefit of sustainable development, the developer is not going to make as much money.

If you are already a green builder, developing your own projects is an excellent way to expand your business. Developers work in many ways. Some of them concentrate on large commercial deals, such as shopping centers. Creating a large subdivision of building lots for homes can prove very lucrative. Small developers can do well by subdividing small parcels of land into just a couple of building lots. Whether you are cutting a single parcel into two building lots or developing hundreds of acres into a sprawling subdivision, there is plenty of money to be made with land development.

Almost anyone can learn to be a land developer. However, developing the skills to create profitable land deals may take a lot of time and effort. The money from developing does not usually come easily. Many developers fail before they see their first projects completed. Getting financing for land acquisition and development can be extremely difficult for anyone who has not established a successful track record. There are many obstacles between the purchase of raw land and a prosperous completion to a project.

> **PRO POINTER**
>
> If you are planning to seek financing for your development operations, you will need a strong loan package. Invest some time in collecting and organizing all your financial data. Since land developing is often looked upon as risky business, the more professional your loan package is, the better the odds are that you will win approval from a lender.

You don't have to be an expert in any particular field to make money as a land developer. Anyone with reasonable intelligence can carve out a place in the world as a developer. Money and good credit are advantages to an aspiring developer, but with creative partnering a person can rely on others for cash and credit. A feel for real estate is important, and the willingness to work hard, learn daily, and make adjustments as you go along all help a person to become a successful developer.

How Much Money?

How much money does it take to become a land developer? The amount of money needed varies from deal to deal and developer to developer. It's possible for developers to begin their work with very small amounts of money. If a person has good credit, the need for cash on hand is greatly reduced. Some

> **PRO POINTER**
>
> You don't have to be a licensed real-estate broker to become a developer. However, building a strong network of licensed brokers will be of great benefit to you.

developers are able to create projects with virtually none of their own cash. Some money is usually needed, but the amount can vary considerably.

As a builder and developer, I have done deals of all sorts. I've taken 200 acres of land and turned it into 10 20-acre lots. Working with a partner, I participated in the development of a shopping center, a large housing tract, and similar deals. On a smaller scale, I have purchased small pieces of acreage and cut them in half to create two building lots. My past experience includes working with other developers to control dozens of lots in a particular subdivision. Most of my experience has been with residential developments.

> **PRO POINTER**
> Before you try to develop land, develop a list of experts with whom you can consult. Some of the types of experts you may need include the following:
> • Soil engineers
> • Surveyors
> • Real-estate attorneys
> • Real-estate brokers

In addition to being a builder, I also owned a real-estate brokerage. During my work as a broker I've helped investors and developers structure numerous land deals. Some projects that look great on the surface turn out to have severe development problems. It's also common for an unused piece of property to have a higher and better use. This is what land developing is all about--finding a higher and better use.

During my career I've seen deals done with as little as $100 from a developer's personal bank account. Developers should plan on needing much more than that, but the out-of-pocket expense can be kept to a minimum. Developers who don't have cash or credit can turn to partners to make a project come together. Almost anyone with determination can find a way to develop real estate.

Large development plans can require thousands upon thousands of dollars before any hope of creating income can be seen. It's not unusual for tens of thousands of dollars to be spent before the first shovel is filled with dirt

> **PRO POINTER**
> Be very careful when taking on partners. Two of my biggest losses in my career have been a result of partnerships. You can do more with partners than you can do on your own, but be sure that you have rock-solid contract arrangements with any partners you choose to work with.

to break ground. Developers working on a plan for a shopping center might spend $100,000 just to get approval for a project. On the other hand, a builder who is splitting one building lot into two lots might not need any cash before financing can be arranged.

Builders with good track records can often include the cost of land acquisition in their construction loans. This is the way I've normally done business over the years. If you create a good relationship with a lender, you may be able to start developing land with the only money out of your pocket being an earnest-money contract deposit to buy the land.

How Do I Find Partners?

How do I find partners if I don't have enough money to work on my own? Potential partners for land developers are everywhere. Some of the partners are investors. If you have a good plan, an investor may put up the money and credit to get the job done. Of course, there will be a need for the investors to see profits. The investors may want a percentage of all profits made, or they may want to earn interest on the money that they are putting up. But money investors are not the only way to get what you need.

There are many types of professionals involved in land-development deals. Surveyors, engineers, architects, and others may be needed in the planning stages of a development. These professionals are sometimes willing to perform their services without immediate payment. The pros might be willing to gamble their time and skills to get a piece of the action. Even banks may be willing to participate as partners with land developers.

Not all homebuilders are developers, but any builder who builds in volume may make a great partner for a land developer. Builders need land to build on. If the builders can turn extra profits by working with a developer, they may be willing to lend their reputation, cash, and credit to create a new development.

Finding partners for land deals is not difficult. However, developers must be careful when choosing partners.

> **PRO POINTER**
>
> Look into getting options to buy lots in new subdivisions if you are a homebuilder who needs an inventory of lots to work with while developing your own land.

And it makes the most sense to limit the partnership to one particular development. Partnerships often end on a bad note. It's usually best when the partnership is limited to a single deal so that later deals can be done together but are not required to be.

More Control

Money is not the only reason for homebuilders to expand into land development. Builders can gain much more control over their businesses when they develop their own building lots. A builder who is not a developer has to buy building lots where they can be found and for whatever price it takes to get them. But builders who create their own developments have unlimited access to the lots that they are developing.

> **PRO POINTER**
> Risk management is crucial to enduring success as a developer. A lot of people assume that small projects carry small risk. This is not always the case. Risk is risk; it is not always tied to the size of a project or the potential profit of a deal.

It's not uncommon for a developer to option or sell large blocks of building lots to selected builders. A subdivision of a hundred lots might be limited to just four builders. This kind of deal is common. Selling to just a handful of builders is convenient for developers. It is also cost-effective in terms of marketing. The downside for builders who are not invited to buy into a development is essentially being locked out of the subdivision.

Some developers limit the number of builders allowed to build in a subdivision to maintain quality construction. It could be that each builder will offer four to six model homes that can be built in the subdivision. If there are 100 lots and four builders who will build four to six types of houses, this works out to 16 to 24 house styles. This type of practice is very controlling, but it goes on all the time. And it can shut out small builders who can't afford to build in high volume.

Successful developers generally have a following of builders. It can be very difficult for a new builder to break into the loop. But if the new builder can develop building lots personally, the heat is off. Builders who do their own developing have much more control over where and what they will build, and they normally make a handsome profit on the land.

How Much Risk?

How much risk is involved in developing land? There can be substantial risk, but it can be managed. Some types of developments create more risk than others. Assuming that all zoning and local code requirements are in order, doing a subdivision of a single lot shouldn't be very risky. Planning to develop a shopping mall is a very different matter. Large commercial ventures can be extremely risky, as can large residential developments.

Many investors feel that risk is directly in proportion to potential reward. Sometimes this is true, but not always. A lot of money can be made without exposing yourself to excessive risk. Risk management is crucial to enduring success as a developer. A lot of people assume that small projects carry small risk. This is not always the case. Risk is risk; it is not always tied to the size of a project or the potential profit of a deal.

There is risk involved with developing land. Of course, there is risk in all types of business. Builders who build on speculation take risks all the time. There can be a lot more risk involved with opening a restaurant than there is with developing land. Basically, the skill, experience, and knowledge of the developer curves the risk factors. Inexperienced people who jump in headfirst and don't take the time to learn what to do beforehand are at much higher risk than seasoned developers. Fortunately, learning to reduce the risk of land developing is not all that difficult. With the right research, you can turn the odds in your favor quickly.

How Can I Compete?

How can I compete with the big guns of the business? Don't established developers have their pick of all the prime property? Developers who have been in the business have certain advantages, but they don't hold all the cards, especially when homebuilders are turning into developers. Many business owners become lazy once they have enough business to satisfy themselves. This can leave a door wide open for aggressive new developers. As for prime property, new developers who dig deeply can find ideal properties before they are offered for public sale. Yes, the big guns will be looking for the same treasure, but it is a race against time, and anyone can win it.

You don't have to possess a long line of credentials to get into land development. If you choose to break into big-time development, you may well be biting off more than you can chew. But there are plenty of opportunities for small investors. Little land deals can turn big bucks, so they should not be overlooked. In fact, this is one reason why new developers have a fair shake. Once developers grow to large proportions, they probably will not pay much attention to a piece of property that can be cut into four building lots. It's not that there isn't money to be made or that the money couldn't be very good. Usually, it's just because the big developers are too busy on major projects to ferret out small deals.

Developers who are not builders have to make all their money from the sale of building lots. There is plenty of money to be made in this way, but builders hold an advantage. Builders can profit from both the markup on building lots developed and on any homes sold. This, in essence, gives builders an added edge of security. If the profits from the land are not as strong as projected, the profit from the houses can still make a project fly. Having the dual earning ability definitely gives builders an advantage over standard land developers.

Red Tape

Isn't there a lot of red tape involved with land development? There can be. Many state, federal, and local agencies may have a say in whether or not a piece of land can be developed for a specific purpose. Cutting through the red tape can take years for some projects, but other projects can zoom through the approval process without much fuss. The work required to gain approval for development is worthwhile for projects with strong potential, but it can be a bank-breaker on marginal deals. Learning to estimate the amount of red tape to be cut is just part of becoming a successful developer.

> **PRO POINTER**
>
> If you are a pure developer and not a homebuilder, consider working with a builder on an exclusive basis. This can make your lots easier to sell, and you should be able to structure a deal where you reap some reward from the builder's profit.

What type of trouble can a developer run into from various agencies? Depending upon where land is located and what the intention for its development is, the obstacles in front of a developer vary. You might have trouble getting approval from a zoning board, but this is a risk that should be set aside as a contingency in the purchase contract for the property. If zoning regulations will not allow the intended use, you should have a clause in your contract that allows you to void the purchase agreement. Be warned: This is not standard language in most boilerplate contracts. You, your broker, or your attorney must make provisions for it.

Environmental agencies can wreak havoc on development plans. Usually, intervention from these agencies is well warranted. However, stopping the development of a large tract of land because there is a frog pond in the back corner may be excessive. Normally, developers can find a way to work with environmental issues so that everyone wins, but it can require leaving a chunk of real estate that was originally slated for profitable development untouched.

Red tape reaches into deed restrictions and covenants, too. A previous developer may have created rules that are not easily broken. Politics can be a factor in developments, as can noise. A developer who is required to install a noise barrier that was not planned for can see profits shrink quickly. The cost of connection to municipal utilities can cut heavily into development profits. Plenty of research is needed to avoid costly problems, but this should not be a deterrent. If a project shows viable potential, the time spent will be well rewarded.

The Transition from Builder to Developer

The transition from homebuilder to land developer is a natural one. Builders who build in volume can benefit greatly from doing their own development work. You don't have to be building 60 homes a year, like I used to, in order to justify land development. Being a developer is enough in itself, but when you factor in the builder's profit, the deal gets sweeter. Most experienced builders have enough industry knowledge to step into the role of a developer without much trouble.

If you are a builder, getting into land development shouldn't be too difficult, and it could prove to be extremely profitable. Are you working a full-time job and wishing you had a business of your own? Would you like to work part-time and make a full-time income? People who hold down day jobs can be developers in their spare time. That's right--you don't have to be available full-time to make major money as a developer. Nights and weekends can provide plenty of time for you to act as a land developer.

CHAPTER TWO

Using Green Space to Make Big Bucks

Using green space to make big bucks is good for the environment and your bank account. Sustainable developments are growing in demand. For developers who want to stay in step with changing times, going green makes a lot of sense. What is the ruckus all about? People are tired of being wasteful. While they want comfortable lives, they want them to be designed and built in a way that has less impact on the environment. And, they are willing to pay for what they want.

I have been involved in real estate and construction for over 30 years. In my younger years it was fashionable to use every available inch of land for salable structures. The mentality was "pave it all.". Fortunately, this has changed. I have seen this change evolving for years, and now we are in a major push towards eco-friendly developments.

Old-school developers may have trouble believing that they can make more money by making fewer building lots. On the surface, the concept is not comprehensible. However, you really can make as much or more by using less land. This gives you the best of both worlds. You get paid handsomely for your efforts, and you contribute to the future of the environment for generations to come at the same time.

How is it possible to make more by doing less? Simply put, people are willing to pay a premium price for quality living conditions. This is not so difficult to understand. Our world has changed greatly over the last few decades. Think about it. If you are old enough, you must remember some key changes that are undeniable. One trend is green developments. To make a point of this, let's step back in history for a moment to share some memories of the way the situation used to be and how it has changed:

- Kids could play tag and dodgeball as a part of school activities.
- Smoking was considered a social event.
- Unleaded gas did not exist.
- Terrorism wasn't talked about daily.
- Water quality in rivers was often ignored.
- Good gas mileage was 18 miles per gallon.
- Electric cars were found only on toy racetracks.
- When you thought of green, you thought of a Muppet frog.
- A calculator was thought of as a computer gadget, and slide rules ruled the engineering world.

Yes, things have changed, and they continue to change. If you don't change with them, you could become as forgotten as leaded gasoline. It is time to go green.

Why Will People Pay More?

Why will people pay more to go green? I think it is a personal decision. Lower impact on the environment is valuable to many people. For this group, paying more for a comfortable community is well worth the price. Different people have different reasons, but they have one thing in common. They are willing to pay more for sustainable living conditions.

Which Types of Features?

Which types of features are consumers looking for in green developments? Many components can come into play when designing a sustainable development. Let's look at a list of some of the key elements of sustainable communities:

- Open space for all to enjoy
- Compact communities that create more green space
- Mixed-use developments
- Walkable communities

- Bike trails
- Exercise trails
- Wildlife viewing areas
- Wildflower sections
- Abundant trees, flora, and fauna
- Underground utilities
- Prime planning for building locations

How Much Am I Going to Lose?

How much money am I going to lose by leaving a lot of land in a natural state? There are differing views on this issue. Exact profits and losses must be calculated on an individual basis. There are, however, two schools of thought on the topic, and those are what we will concentrate on here.

Some people say you can't squeeze blood from a turnip, and there are land developers who will do their best to mimic the old saying. The mindset is that you have to hack and slash trees to make room for more housing and commercial developments. This used to be common and still continues. But the public is speaking. And it wants change. Green developments are the wave of the near future. You could lose much more by refusing to create sustainable development than you would by creating green space in your next development.

Many people feel that developers can actually develop green projects with less cost than a traditional development. Why is this? The thinking is that money is saved by doing less site work and leaving more natural area. The cost saving is real. But what about the loss of buildable land? Accounting for the reduced building space can be difficult. Is it a toss-up? Any way that you look at it, the world and future generations win with green developments. If nothing else, this is added value.

Get in Early

To cash in on the green craze, you should strive to get into the game early. In business there is risk and reward. Much of the risk of going green has already been taken. The path has been proven to be a safe one for what appears to be many years to come. That leaves the reward.

Depending upon your age and how long you plan to be active in land development, sustainable development is likely to become a substantial part of your career. Don't overlook the opportunity. You have taken a big first step in reading this book. Keep it up. Learn the benefits and advantages of eco-friendly developments to enhance your bottom line.

CHAPTER THREE

Build a Strong Development Team

A lot of people wonder if they have what it takes to break into land development. Many of these people write themselves off as being unsuited to the business. In some cases they may be, but you might be surprised at how little it takes to get a piece of the action in land developing. You don't have to be an engineer or a licensed real-estate broker. Developers are not always builders, so you don't have to know how to hammer nails. In fact, there are no real minimum qualifications to meet on an official level. This is not to say that developing land is easy. But you can train yourself to be a developer and start making money as quickly as you are ready to.

Real-estate investors don't need special licenses. If they have money or credit, they can buy and sell real estate. The brokers with whom they may work are required to be licensed, but the investors are not. Neither are developers. In fact, developers are real-estate investors. But rather than just buying and selling, developers change the condition of a property. They don't always build on it. Sometimes they acquire property, get it zoned for development, and then sell it for a hefty profit. There are many variations in ways to work as a developer.

Becoming a successful land developer is not always easy. In fact, it rarely is. But it can be much easier than earning a college degree, and it can prove to be much more profitable. Even with a degree, graduates usually have to work for years to repay their college loans and to climb the financial ladder to their prime income potential. A developer can jump into a six-figure income in a matter of months. But, just like the college student, wise investors and developers spend hours reading and doing research to ensure their success.

Who can become a land developer? Almost any adult can delve into development. The background of some people is better suited to the business than others. You will have an advantage if you have experience in any of the following:

- Real-estate sales
- Homebuilding
- Excavation
- Engineering
- Surveying
- Real-estate law

No special skills are required to become a land developer. Anyone with reasonable intelligence and a desire to become a developer can approach a venture with optimism.

Ideas

Ideas are one of the most important tools in a developer's arsenal. Taking a piece of land and cutting it into four equal pieces qualifies as land developing. This type of project can produce good profits and does not require much creative thought. Planning a full subdivision is a very different job. Few developers create their own development designs. Usually landscape architects and site planners come up with the designs. But developers who are involved in their own designs are often happier with their projects, and these projects are sometimes more profitable.

It is often assumed that developers don't need design skills. This stems from the fact that experts are usually retained for such work. While it is true that experts often do the actual design, good developers have to have an eye for design to see the potential of a prospective property. If you have no creative ability to see something that does not yet exist, you may have trouble as a developer. Successful developers can look at a tract of raw land and see glowing images of what could be built on it. Call these people dreamers if you like, but they can see visions and then move forward to make the dreams come true. This is a natural talent for some developers and a learned skill for others.

Money and Credit

As I have said previously, you don't need a lot of money or credit to become involved with land developing. It certainly helps to have both, but determined developers can find ways to overcome a lack of cash or good credit. Doing a partnership deal is often the easy way to circumvent problems of limited resources. Negotiating with the seller of the land for purchase terms is also an option.

> **PRO POINTER**
>
> New developers may not have an eye for design features. This is a skill that will come with experience. However, you must be aware of the benefit in seeing land with a creative slant. Once you begin looking for special features, you will learn how to see them.

Some sellers are quite willing to offer very agreeable terms to developers. In fact, there are sellers who are happy to participate as partners with developers. Raw land can be difficult to sell. This can make sellers more receptive to creative financing options. This is especially true if you are able to spot a piece of land that is not easily identified as a desirable property. Most sellers are willing to accept small deposits, and they don't often run credit checks on buyers. Let me give you a few examples of how you might involve the seller of land in your development plans when you don't have much money or good credit.

Seller Financing

Seller financing is one way to involve a seller in your development plans. A lot of land is sold with seller financing. All financing agreements should be reviewed very carefully. In the case of seller financing, an attorney should review all terms of the loan agreement

> **PRO POINTER**
>
> Never overlook the possibility of involving a landowner in your developing plans to keep your out-of-pocket expenses down.

before any final commitment is made. When you can arrange good seller financing, it can make it possible to get a jump ahead in the development game.

If a seller is willing to work with you and is in a position to offer seller financing, you can do just about any type of land deal that you can agree to with the seller. A tiny deposit could give you contractual control over a nice parcel of land. Once you have the land secured, you can move on with your preliminary development work. It can take months, sometimes years, to get all approvals needed for a development plan. A seller may be willing to hold a piece of land for you during this period of time, and you might not have to make monthly payments. The seller might agree to let the payment amount accrue until you gain development approval.

It's fairly easy to get interest-only loans from sellers. This type of loan is advantageous due to lower monthly payments. You might have to offer the seller a higher sales price to get more attractive terms. The added cost of the land may be a small price to pay in order to get attractive financing. Buyers and sellers can create almost any type of deal between themselves.

Options

Options are another consideration when dealing with a seller. It's possible to control vast amounts of real estate with nearly no money when you use options effectively. A $50 bill can give you control of hundreds of acres of land. The problem with options is that the money that is given as option money is usually forfeited if the optioned property is not purchased. But the option money can sometimes be kept low. An advantage of options is that the option money is all that is at risk, so the pendulum swings both ways.

Seller Participation

Seller participation in development deals is certainly not unheard of. When you can work directly with a seller, you may have any number of opportunities. Some sellers may be very happy to work with a developer as a partner. When this is the case, the developer doesn't have to buy the land. Avoiding the cash outlay to purchase property may be enough of an advantage to bring a seller into a partner position. If you have a solid development plan, you may even find a seller who will help you pay the predevelopment costs in return for a percentage of the overall profits. It never hurts to ask a seller for some form of participation in a deal. The worst that can happen is that your request will be denied.

Organizational Skills

How are your organizational skills? Being well organized is a major plus for a land developer. If you have trouble remembering appointments, getting to work on time, or finding personal papers when they are needed, you could be in for a rough ride as a developer. Good developers are well organized and meticulous in record-keeping matters. Of the many skills needed to be successful as a developer, strong organizational abilities are certainly high on the list of top priorities.

Getting organized is difficult for some people. But, it's not terribly hard to gain satisfactory organizational skills. If you are not already comfortable with your ability to file and find material, put some effort into learning how to manage your business matters. Once you start developing land, there will be plenty of material to keep up with. You can improve organizational skills with some of the following ideas:

- Read books on time-management and organizational skills.
- Attend seminars on organizational skills.
- Use trial-and-error techniques to see what works for you.
- If you are not already working with computers, learn how computers can organize your business efforts.
- Set up a filing system that you understand and are willing to use.

Research

Research plays a vital role in land developing. It is often deep research that turns up outstanding deals. Rarely is a prime deal advertised for sale in a newspaper. Real-estate brokerages are a source of land for sale, but, like ads in a newspaper, the deals offered by brokerages are probably not the best deals available. Savvy investors ferret out their own deals. Here are some ways of finding good deals:

- Research tax records to establish property values and owners.
- Ride around to find suitable properties and then look up the owners in public records.
- Network with local people who may know of something that is coming to market soon.
- Lawyers are a good source of hidden deals. When estates have to be settled, there can be land for sale. It's common for lawyers to handle the settling of estates, so getting to know local lawyers is helpful.
- Talk to bankers. They often know of land that has been foreclosed on and that will be coming up for sale.

People Skills

Having good people skills is very helpful for a land developer. There are many people who must be dealt with during most development projects. If you are uncomfortable talking to people, you are going to find yourself at a disadvantage. Lacking general communication skills is a detriment in most careers, and it can be a severe hindrance to developers.

PRO POINTER The best buys on raw land are not usually found in newspaper ads or even from average real-estate brokerages. Finding the best deals takes personal research and effort.

Someone has to talk to sellers, brokers, builders, buyers, surveyors, engineers, code officers, and a lot of other people. If you can't manage this, you may have to consider hiring someone to do your talking for you. This can be done. But it's better to work on your own abilities and rise to a level of comfortable communication on your own.

Time

How much time must be devoted to a development? The amount of time needed for a development depends on the nature of the project and the skill of the developer. Simple development plans may be ready to work with in just a couple of months. Complex developments can require years of preliminary work before the land can be developed. How much time must you have to devote to developing on a regular basis? Again, the answer is elusive for many reasons. However, small development deals can generally be managed by people who have full-time jobs. It's very possible to get into land developing on a part-time basis.

If you can't make yourself available during normal business hours of a week, you will be at a disadvantage. At the very least, you should try to be available for phone calls. Some parts of the preliminary developing work can be done after regular business hours. Brokers, sellers, subcontractors, and others are often willing to work evenings and weekends. You are not likely to find

PRO POINTER If you have difficulty keeping appointments, being on time, and getting work done in the time allotted, you should put effort into learning time management. Time is a big asset, so learn how to capitalize on it.

engineers and code officers so accommodating. It is difficult to do all aspects of land developing if you don't have some flexibility in your schedule for appointments during normal business hours.

The amount of time needed for a development is very difficult to predict. You may find one project that will almost run itself and then hit a deal that requires your attention almost constantly. If you will be working with a real-estate broker, you may be able to use the broker for some of the daytime requirements of meetings and deliveries. It's fairly easy to enter developing on a part-time basis, but flexibility during the day can be a big factor in how quickly your projects come to life.

Much of the work done during predevelopment tasks is done by people other than the developer. Once the development team is in place and working, the requirements for a developer diminish. You can limit the amount of time required of you if you build a good team of professionals to work with.

> **PRO POINTER**
> Having a flexible schedule that allows you to attend meetings during normal business hours is an important element to successful land developing.

Choosing Your Development Team

Choosing your development team is a task that you should take on as soon as possible. Even if you don't have any idea of the exact property that you will be developing, you should start to assemble your team right away. This is a process that takes time, and it's such an important element that you should never rush it. Depending upon what your basic plans for development are, you may need a large group of professionals to get your project off the ground.

It's easy enough to open a phone directory and hire a surveying firm, but finding a firm that you are comfortable with and that has the skills you need could take weeks or more. Large projects require a diversified crew. Some of the players needed on your team may include any of the following:

- Land planners
- Architects
- Landscaping architects
- Engineers
- Designers

- Drafters
- Tax attorneys
- Accountants
- Marketing experts
- Project managers
- Soil scientists
- Environmental experts
- Geologists
- Hydrologists
- Real-estate brokers
- Builders
- Various consultants

Small projects don't usually require such extensive talent. But there will be a need for experts, and you should find them long before you need them.

If you are new to developing, you may not realize how important it is to have a good relationship with the members of your development team. Working with people you like and can communicate well with makes every project more enjoyable.

The Experts

The experts needed for your team will depend on what you are developing. A simple subdivision of a building lot will not normally require a lot of technical attention. If you have plans to develop a large subdivision, you can expect to need a lot of experts.

Another factor is the location in which you will be doing your development. Some regions require different types of talent. Code and zoning requirements can also affect the type of experts required for a project. Depending upon your skills, you may fill some of the slots that would otherwise be held by experts. For example, you may be able to act as your own project manager or as

> **PRO POINTER**
> Communication is always a key to business success. Put people on your team who can and will communicate clearly with you. Never accept an expert whom you can't understand.

your own marketing consultant. Let's look at some of the experts that might be needed on your project.

Attorneys

Attorneys are almost always needed in some capacity for a development deal. You may consult with tax attorneys prior to making a decision to do a development. A good real-estate attorney should be involved in contract issues and the settlement procedures of purchasing property. If you are unfortunate enough to run into litigation, you may need an attorney to represent you in court. Don't expect a single attorney to fulfill all your needs. It's better to deal with specialists in each field.

Accountants

Accountants are often consulted prior to major developments. Certified Public Accountants (CPAs) are the standard choice for expert tax advice. Most developers depend on their CPAs to stamp their approval on a project before much effort is put into a development. When partnerships are created, accountants may be used to render accountings to each partner on a regular basis.

Marketing Consultants

Marketing consultants may be used to determine the viability of a project. The consultants can range from real-estate agents to general marketing consultants. Many developers learn to be their own marketing experts. They do this to have an increased comfort level in the proposed success of their projects. Small projects don't generally yield enough profit to justify a large budget for intricate market studies and sales plans. On these types of projects builders, brokers, and developers put their heads together and come up with figures. Large projects may not be approved for financing without a formal market study. These studies can be very expensive; don't forget to budget for this expense if you have any reason to believe it will be needed.

Builders

If you will not be doing your own building, you may have to talk to a number of successful homebuilders during your planning stage. The builders should be able to give you an idea of the types and price ranges of homes that they might build in your development. As a rule of thumb, land value is usually thought of as being worth about 20 percent of the total appraised value of a home. A house that sells for $200,000 could be built on a lot that sold for $40,000. If you are hoping to get $50,000 for each of your building lots and find that the builders would be building houses that didn't exceed $175,000, you would be hard pressed to get your price for the building lots. This could kill your deal. Of course, your marketing consultants should be able to help you with this aspect of your planning, too.

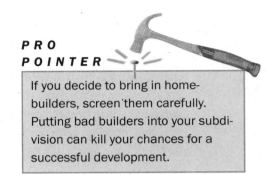

PRO POINTER

If you decide to bring in homebuilders, screen them carefully. Putting bad builders into your subdivision can kill your chances for a successful development.

Project Planner

You may be your own project planner, or you may hire one. If you do contract with or hire a planner, you should have your arrangements in writing. The planner bears a lot of responsibility. If you hope to be your own planner, make darn sure that you are competent to get the job done properly. If you hire an outside planner, choose one who has a long, successful track record. Project planners pull in data from other team members and then put it all together in a master plan. If you are thinking in terms of homebuilding, you could compare a project planner's role to that of a general contractor.

Architects

Architects are often instrumental in the design stage of a development. Landscape architects play an important role on projects that will require extensive landscaping. However, cutting up rural land into building lots for a few houses will not likely justify the services of a landscape architect. If your development plans include building a structure, architectural plans will probably be needed to obtain full development approvals. As with all your experts, written agreements should be in place for the services rendered and the payment offered.

Civil Engineers

Civil engineers are frequently involved in development projects. The engineers may work with geotechnical conditions and infrastructure requirements. It is fairly common to find civil engineers working in firms with other types of talent that will be needed for developing land. For example, a firm that I used to use could handle all of my engineering, drafting, and surveying needs. It helps to keep as much of the work as you can under one roof. Have your attorney draft a working agreement for all of your dealings with engineers.

Surveyors

Surveyors are needed for practically every development deal. The extent of work done by the surveyors varies. You might need a simple boundary survey, or you might need elevation surveys. Normally, surveyors are on a project several times before their job is complete.

Finding the Right People

Finding the right people for your project can be frustrating. If you are not connected to the developing industry, you will have to start from ground zero. Builders have a slight advantage in that they often come into contact with various experts during their building chores. A quick skim of a phone directory will get you started, but hopefully you can find a better way to find the best people with less effort. If you know people who work with the types of experts that you will need, talk to your acquaintances. Word-of-mouth referrals are generally the best way to find good talent. If you don't have many connections in your community, consider joining some clubs. A lot of networking goes on during club meetings.

> **PRO POINTER**
> Take the time to meet with potential experts. Face-to-face meetings are very important in establishing how you feel about working with someone.

It's been my personal experience that people in the building and developing businesses are willing to share information about people whom they have done business with. Asking a fellow builder for a lead on whom you should talk to for a fast soils test is not threatening. If you were asking about the builder's plans for next year's home designs, you'd probably get stared down, but this is unlikely to happen when you are asking for leads on development talent. Unlike subcontractors, whose names builders may keep quiet about to keep from losing them, the experts used for developing land are in a different category. Joining a local builders' association can be a fast way to get some good information about people to talk to.

When you make a good contact, work it. For example, assume that you already have a CPA whom you are comfortable with. Ask that CPA for the names of some good tax attorneys. Use every contact to create another contact. The snowball effect of this procedure can fill your phone files quickly with names and numbers. If worst comes to worst and you are forced to cold-call from a phone book, take the time to meet with potential experts. Face-to-face meetings are very important in establishing how you feel about working with someone.

Questions to Ask

When you are meeting with experts to interview them for your project, be prepared with a list of questions to ask. You may not be given a lot of time to talk with the experts, so be well prepared. Don't be intimidated by them. Remember, they will be working for you. Some rookie developers go into relationships with experts and feel in awe of them. The experts should be respected, but the developer is the one in control and signing the checks. Don't forget this. These are some of the questions you should ask:

- Ask for their credentials.
- Inquire about previous projects that they have done.
- Ask for a list of their clients. If they won't consider this request, at the very least get some references.
- Ask about projects that are presently in development that you can see.
- Ask about fees and terms.

A Rough Plan

Once you have found a group of experts whom you believe will be comfortable to work with, you should come up with a rough plan. I'm not talking about a site plan or blueprint. The plan that I'm referring to is your preference for a project to undertake. How big will the project be? Are you going to do something within a city or out in the country? Is it your intention to develop a piece of land and then sell it as a whole project, or will you cut it up and sell it in parcels? Start asking yourself these types of questions.

Once you begin to flesh out your overall intent for a project, you will know more about which experts will be needed. For example, if you will be developing land where private sewage-disposal systems are required, you will need someone to do soil testing. This type of testing will not be needed if you are going to develop in an area where city sewer hookups are available. Define your rough plan as much as you can, and then figure out which types of experts will be needed.

Once you have a list of experts and a good idea of the type of project that you plan to do, it's time to make some phone calls to your experts. Explain to them what your thoughts are. Give them a brief rundown of how you see things, and ask for their advice. Find out how much time is likely to be needed for each expert to complete the work required on such a project. This will allow you to begin building a framework of what to expect when you find the land that you wish to develop. Confirm, as best you can, that you have covered all the bases in terms of the list of experts needed. Not all experts will be willing to help you double-check yourself, but some of them will. This aspect of the planning process will give you a fair example of how each of your experts will be to work with when the real project comes in.

> **PRO POINTER**
>
> Proper planning is crucial to a successful development. Take enough time with your planning to explore all viable options. This is often when you will find the greatest value. What you are looking for is the highest and best use for the property that you plan to develop.

Lenders

Lenders are almost always used in development deals. Most developers must borrow money to do their developing work. Many developers wait until they find a piece of land before talking to a banker. It's better if you have already established a relationship with a lender when you find land to purchase. I strongly suggest that you get out and meet lenders well in advance of needing money. Wait until you have your list of experts to take with you. If you have chosen well, the names of your experts can help put a lender's mind at ease. Knowing that seasoned, successful professionals will be working with you may make a banker a bit more willing to do business with you.

Not all lenders offer loans for land development. Many lenders consider the business too risky. Call around and see which lenders in your area are willing to talk about loans for land developing. Make appointments to meet with various loan officers. Take a current balance sheet in with you when you meet with your lenders. It makes sense to start with lenders whom you already do business with, but don't limit yourself to your favorite bank. Get out and meet as many loan officers as time will allow for. Once you start rolling, you will probably want to borrow from more than one bank at a time.

Don't expect loan officers to be too quick to give you answers about the odds of a loan approval. Most lenders consider each deal on an individual basis. A developer's personal credit, assets, and reputation are important when borrowing money for a development, but there's more to it than just that. Good development plans sell themselves to bankers. If a development can carry itself strongly on paper, less importance is placed on the financial worth of a developer. Therefore, lenders are reluctant to preapprove any loan without full facts.

The goal in the early stages is to get to know the lenders and to let them become comfortable with you. This will help immensely when you make a formal application for a loan. By meeting with various lenders you will also determine which ones seem the easiest to get along with. Developers usually spend a good deal of time working with lenders, so it's important to have a good working relationship. Once you are comfortable with your team, you can begin to search for land.

CHAPTER FOUR

Making Sales

Creating a marketing plan is a big part of land developing. You may find it beneficial to create your own sales team, or you may choose to list your property with a real-estate brokerage. In one way or another you will need a plan for making sales. After all, what is a good development worth if you can't sell it? Seriously, having a strong marketing plan and sales force is most likely what will make your deal profitable. There are many sayings in the business, and some of them ring true. It's common to be told that good marketing can sell a bad development, but that bad marketing can't sell a good development. This is one of the true statements. Sales are sales, and sales are what make the developing process profitable.

Every developer needs a solid sales plan. The type of plan can vary considerably from project to project. For example, selling building lots will require different talent than selling spec houses. Leasing professional space in a new development requires another type of talent. You have to look at your project and form your sales approach accordingly.

Some developers concentrate heavily on sales, but many are so consumed in their development work that they don't spend a lot of time on sales. Some developers make their own sales, but most rely on others to close deals. You can use either a professional brokerage or an in-house sales force. Both methods can be effective.

As a builder and developer I have used real-estate professionals and my own staff to make sales. Most of my success has come from an in-house sales team. However, I have had brokerages that served me well. There are pros and cons to both approaches. When I became a developer, I knew little about sales, but I took the time to learn the ropes, and it paid off. You may not be making your own sales, but it doesn't hurt to know the business.

If you are doing a small development, listing your property with a brokerage is probably your best bet if you don't feel that you can sell it yourself. Larger projects can easily justify an in-house sales staff. The commission rates for sales can vary greatly, but they are usually comparable. There is no rule that sets a commission rate. In fact, it is illegal to fix commission rates. Being your own salesperson is the least expensive route, but the other options may be better for you. Let's spend a little time considering how you might make the most money from your development.

Independent Sales

Independent sales are good in that you don't have to maintain an overhead expense for employees. While this is good on one hand, it can be bad on the other. Choosing between an employee-based sales force and a commission-based sales team can be tough. Simple math will tell you how much each type of sales attack will cost. Here are some questions and facts to consider:

- Is it better for you to pay only for what is sold?
- Could you make more money by having an in-house sales team?
- Listing your property with a professional brokerage doesn't cost you a cent until you get a closed sale.
- Many developers who have medium-to-large projects maintain an in-house sales staff.
- More often than not, however, developers list their properties with outside agencies.
- How much are you going to have to sell?
- How many sales will it take to sell most of your development?
- Can you afford the overhead expense of advertising and salaries until sales close?

There can be a multitude of questions to ask prior to making a firm decision. There is no cookie-cutter answer. Each developer and each project have specific needs. With this in mind, let's look at the option of using an independent brokerage to sell your development.

Brokerage Sales

What is the commission rate that you will be charged on brokerage sales? Independent brokerages usually charge their clients a percentage of the price paid by a buyer at a closing for their services. The percentage of the price is called a sales commission, and the precise percentage can vary considerably. Commission rates are not fixed, and it is illegal to fix them. However, they do tend to be very similar from one brokerage to another. Commission rates are negotiable. It is illegal to fix the rate, so it is always negotiable. A brokerage that is paid an 8-percent commission on one deal might do a different deal for a 6-percent commission. Some sales agents try to convince sellers that commission rates have to be a certain amount. This is not true. A brokerage may decide that a minimum rate is required to make the sales effort worthwhile, but you can always ask for a lower rate.

> **PRO POINTER**
> I believe in keeping the reward for a brokerage high enough to stimulate fast sales.

There is a downside to cutting a brokerage's commission rate. If you don't make it attractive for a brokerage to sell a property, you may not see a lot of sales activity. When a brokerage has a large commission at stake, it can be a great motivator for making a sale. Having been on all sides of this table, as a builder, a developer, and a brokerage owner, I've seen the angles from all perspectives.

When you first look at the amount of a commission, it can seem very high. Assume that you are selling a new house for $200,000 and paying a 7-percent commission. The amount that you would pay in commission would be $14,000. This is a lot of money. But what will the brokerage have to do to earn it? There will be advertising, and that can be costly. Usually there is a lot of wasted time working with people who don't buy. Time, effort, and expense have to come out of the commission. It's still a lot of money to pay for one sale. How much would it cost you to sell the house yourself? If you have the ability, you might save several thousands of dollars selling your own properties. I did plenty of this in my early years.

If you are your own salesperson, you can sell for a lower price and still net the same amount of profit. Or you can sell for the market rate and pocket more profit. You clearly have more options and more control. However, you do have the expense of advertising and the burden of showing the property. You have to decide what your skills are and what you are willing to do to limit the cost of commissions.

Paying large commissions can be difficult to swallow. But, having independent, commissioned salespeople moving your property can be a good way to go. If you are not sales-oriented, don't have your own sales team, and simply are not interested in being active in your sales, an independent brokerage should look appealing. The good thing about commissions is that you don't pay them until you have sales. Brokerages take all the risks up front and only get paid if they sell your property.

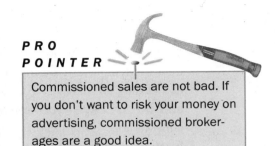

PRO POINTER

Commissioned sales are not bad. If you don't want to risk your money on advertising, commissioned brokerages are a good idea.

What's the bottom line? Negotiate for a fair commission rate, but don't make the figure so low that your brokerage will put your property on a back burner. If you have a large project, consider building an in-house sale team. Selling a spot lot or a spec house here or there should either be done by yourself or a commissioned brokerage. Small deals don't warrant salaried sales staff.

Listing Agreements

There are many types of listing agreements that may be used when you engage an independent brokerage. They are:

- Exclusive listing agreements
- Exclusive agency agreements
- Open listing agreements

If you enter into an exclusive listing, often called an exclusive right to sell agreement, you will have to pay a commission to the listing brokerage even if you sell the property yourself. This is not the case with an exclusive agency agreement. When you have an exclusive agency agreement, you can sell your property yourself and pay no commission. This is the type of listing that I would want to use as a developer. Brokerages don't like having anything less than an exclusive right to sell listings, but they will usually accept an exclusive agency listing. Many brokerages will not accept an open listing, which is essentially a listing where whoever sells a property first, which can include many other brokerages, is entitled to a commission.

It is your right and in fact your duty to ensure that the listing agreements that you enter into will prove beneficial to you. Signing an exclusive listing with an agency is a major commitment. There have been times in my career when a brokerage took a listing and did very little. Your hands can be tied in this situation. You need some control over your listing agreement. If you don't have a right to terminate a listing agreement, you could be left hanging in the wind for months. Don't allow yourself to be put into this type of position. Have clauses in your agreements that allow you to cancel the agreement if the listing brokerage is not living up to the agreement between the two of you. I suggest that you consult your attorney to arrive at the proper wording to insert into listing agreements.

PRO POINTER
If you are going to sell property that is owned by someone else, you are most likely going to need a real-estate license.

Being Your Own Salesperson

Being your own salesperson can take a lot of your time. The expenses of advertising will come out of your pocket if you sell your own properties. I am not aware of any state that requires a developer to be a licensed real-estate professional to sell property owned personally. This means that you should be allowed to sell property that you own without any special license. It does not mean, however, that you can sell spec houses for builders who bought lots from you.

Selling your own property can save you a lot of money. Commission rates add up. If you are good at sales, you might find that you can enhance your profit potential substantially by selling your own projects. Remember, though, that the sales game takes time. If you are able to make sales, the money saved can pay for your time. Deciding to be your own sales force is a big commitment.

Your Sales Team

Having your own sales team makes a lot of sense if you have enough to sell. Run the numbers and look at them for yourself. Assume that you have a 12-house development. By most standards this is a small development, but it can be large enough to justify your own sales team. Assume that the sales commission on each house will be $20,000. With 12 houses this works out to $240,000 in sales money.

PRO POINTER
Having your own sales team makes a lot of sense if you have enough to sell.

You will need a chunk of it for advertising. If you hire an in-house salesperson, you will have to pay a salary. Once you have an employee, there are other expenses that get tacked onto the basic salary. Let's say that your sale pro is going to cost you $60,000 a year when all costs are factored in. This leaves you a lot of money to use on advertising.

If you are thinking of hiring an in-house sales staff, you should check local laws and regulations to see if the employees will need any special type of licensing. In many cases the employees of a company can sell real estate owned by the company without any special licensing needs. Assuming this is the case, you will only have to deal with typical employee overhead and responsibilities. However, you should see to it that your sales team is trained to avoid legal confrontations.

The sale of real estate can result in lawsuits. Having inexperienced people who are not familiar with real-estate law can be very risky. If you are not up to speed on real-estate law yourself, require your sales team to attend training classes on the subject of real-estate principles and practices. Invest in some errors-and-omissions insurance to protect you and your company if the sales team makes a mistake. The insurance is added overhead, but it is an expense that you should not attempt to operate without. One lawsuit could wipe out your entire company, and the insurance could protect you from such a disaster.

A Marketing Plan

Once you know who will be selling your development, you need a marketing plan. You or your sales experts have to come up with a plan for selling your lots, space, or buildings. Some developments require only simple plans, but other developments can be worth investing heavy money for a marketing plan. You have to weigh your personal situation. Obviously, you would not pay a marketing consultant $45,000 to tell you how to sell one or two houses. But if you were developing a complex of 300 houses, paying a consultant might make a lot of sense.

When you begin to think of marketing plans, you could be considering a number of possibilities, such as:

- To say that you will spend $5,000 or $50,000 on advertising is not enough.
- You have to pinpoint how you will advertise.
- Are you going to rent space on billboards?
- Will direct mail be a part of your marketing plan?

- What success would you expect from television commercials?
- Would radio spots work for you?
- How much are you willing to spend on signs and where will they be placed?
- How soon will you start your marketing campaign?

Your marketing plan will depend a lot on who your buyers will be. For instance, if you are planning to sell blocks of lots to builders, a direct-mail campaign should work nicely. It wouldn't be feasible to use direct mail to general consumers, but if your target market is small and identifiable, such as it would be with homebuilders, direct mail could produce great results. Running radio spots should work for homebuyers, but they wouldn't be worthwhile if you are after builders. You have to match your marketing to your audience.

Let's assume that your primary market is comprised of homebuilders. You might want to tell the builders that you will be running ongoing ads for the development to make homebuyers want to live in it. This could influence builders to buy from you. When this is the case, you should have a defined advertising schedule to show the builders. The schedule should be in writing, and it should be firm. For example, you might promise the builders that you will run three radio spots a day for the first 120 days after the development is ready to be sold. This type of promotion can be expensive, and you might want to avoid it. But if it helps you sell out to builders quickly, the added expense will be well worthwhile.

An entire book could be written on marketing and advertising. In fact, many have. You should read some of them to get a feel for what might work with your development. There is not enough room here to teach you how to build a formidable marketing plan. It is, however, crucial that you be aware of the need for such a plan. You have to establish an advertising budget and a viable means for making people aware of your project. Getting quick sales at good prices is the key to winning in the developing game.

You should turn your attention to a marketing plan early in your development. It is advisable to have your plan completed before you even begin construction. Don't wait until you have something to sell to figure out how to sell it. Concentrate on getting sales before you have a finished product. Presales are always welcome. Go get them as soon as you can. Once you have your sales rolling, you can begin to think about new deals.

CHAPTER FIVE

Setting Up Your Lenders and Investors

Most developers will tell you that financing is the most important part of any deal. It easily can be. Getting even a small development deal off the ground without financial backing is difficult. There are very few people who can pay cash as they go for their development expenses. Most people have to finance their deals, and even developers who could come up with the cash often prefer to use financing. Let's face it, most of us don't pay cash for cars, and we use plastic to buy stuff that we don't really need. Financing is a part of American culture. For developers it can be the difference between success and failure.

It is common for developers to create entire project plans before seeking financing. To get approval for a development loan, a lot of paperwork has to be done. However, you can make some preliminary inquiries before making a formal loan application. And you should consider all potential sources of financing before you cast your net for borrowed dollars. Commercial banks are the most common source of development money, but they are far from the only source. You might find money from a credit union, a private investor, a mortgage broker, a mortgage banker, or even the seller of the land that you are looking to buy. There is also the possibility of using limited partners to generate cash. Builders may participate with you, as might material suppliers. There is potential for a lot of creative deal making with land development.

Where do you plan to look for your development money? Most inexperienced developers go to their regular bank as a starting point. This is not a bad move, but don't take rejection from your bank as a deathblow. Many lenders simply don't like to loan money for land deals, which are generally considered a major risk. You may have to talk to dozens of loan officers to find a few who are willing to gamble on development projects. The time you spend in prospecting lenders is well spent. If you are like most developers, there isn't going to be much activity to brag about until you have a willing lender.

If you are new to the land-developing business, your search for money and favorable loan terms is likely to be more arduous than it would be for a developer who has a successful track record. Lenders are usually nervous about development deals regardless of who is doing them, and if you are trying to do one without any past performance to judge by, you are going to have a tougher time getting financing. But a good deal is a good deal, and lenders who know the land business can recognize a good deal when they see one. With or without a successful history, you can get the money you want if your deal is good enough.

Putting together a winning proposal for prospective lenders can take weeks. Don't think that you can walk into the office of a loan officer, chat for a while, and leave with a loan commitment. You need a written proposal for how you are going to make your development grow and prosper. If the proposal is solid, you have a good chance of gaining loan approval.

Many developers are intimidated by lenders. You shouldn't be. Lenders make their money by lending money to others. You are a lender's customer. The lender is not doing you a favor by loaning you the money you want. In reality, the lender is hoping to turn a profit by getting you to borrow money. This whole concept is difficult for some developers to adjust to. Many borrowers feel that their banks are doing them a favor. The truth is that you are entering into a business deal with your lender. You hope to make money on the development, and the lender hopes to make money on the loan. It should be a reciprocal relationship. Don't bow and scrape. Stand tall and pitch your development as a prize catch. Think of yourself interviewing the lender rather than being interviewed by the lender.

It's difficult for many people to face lenders on equal ground. Most borrowers go into a bank as if they were begging for something. Don't do this. Cast a positive, authoritative view of yourself. Would you feel intimidated talking to subcontractors who are going to haul dirt for you? I doubt it. Think of your lender as just another subcontractor. Yes, you need the money, but you have many lenders to choose from. You are not dependent on a single source of money. Remember this. A positive attitude can take you a long way in the banking world.

Commercial Banks

Commercial banks are the most common source of financing for land development. Are they the best source? They can be and often are. But keep in mind that banks are not the only source of funds for developing projects. Commercial banks lend money for a variety of reasons. Most of them will at least consider land loans and development projects. To get a project approved by a loan committee, you may have to offer substantial paperwork. Much of what a bank will want to see will be reports from your experts.

> **PRO POINTER**
>
> You will have to impress the bank with your ability to complete a project successfully. You have to give your loan officer detailed reports and proposals that will support the likelihood of your success with a particular project.

Loan officers probably won't have much interest in your specific visual plans. They won't care if a house has a roof pitch of 8/12 or 10/12. Most of what the lender will be interested in are the financial solutions. You will need solid sales projections, a sales plan, and reports on what the value of your development will be. Banks do care that developers know their business. You will have to impress the bank with your ability to complete a project successfully. You have to give your loan officer detailed reports and proposals that will support the likelihood of your success with a particular project. Anything less than a convincing proposal is likely to be shot down at the loan committee. Bottom line: You have to sell your development to your lender before you can sell it to buying customers.

Conventional lenders generally like a lot of documentation. For some reason the quantity of paper seems to make a loan request feel safer for most bankers. Give them what they want--that's what I always say. To get a loan from a commercial lender, you have to know what the lender is looking for. This is easy enough to figure out. Make an appointment to meet with the loan officer. When you meet with your banker, ask plenty of questions. Find out what the lender is interested in and then go back to your office and create it.

Talking to lenders is the best way to find out what they want to see in order to approve your loan. I could give you samples of loan applications and loan packages, but they might mean very little to you. It is a much surer bet to ask your particular lenders what they want from you. They will typically want a lot, but it should all be material that you are having prepared for your purposes anyway.

There are certain basics that nearly every lender will want. A typical loan package will include the following:

- Copies of tax returns for the last two years
- A profit-and-loss statement on your business
- A financial statement of your personal worth
- Unless you are a heavy-hitter with strong corporate assets, expect to put your personal assets on the line--very few lenders will accept a loan document signed by you as a corporate officer with no personal liability
- A site plan
- A master plan
- Detail plans
- Sale projections
- Sales procedure

Mortgage Brokers

Mortgage brokers are to money what real-estate brokers are to property. When you go to a mortgage broker, you have a person who puts you, a person in search of financing, in touch with a lender. The broker charges a fee for services rendered. The fee is usually in the form of points on the loan, but it can be structured in various ways. Good mortgage brokers have long lists of lenders for various types of projects. When a borrower comes to a broker, the broker reviews the loan request and then circulates it among perspective lenders. The result can be quick financing, but it can also be a waste of time. Not all mortgage brokers are well connected, and some of their financial sources may not be typical lenders. You have to be careful when dealing with a broker, but you should not rule out this line of financing.

> **PRO POINTER**
> Not all mortgage brokers are well connected, and some of their financial sources may not be typical lenders. You have to be careful when dealing with a broker, but you should not rule out this line of financing.

Money borrowed through a mortgage broker may cost you more than if you borrowed the money directly from a bank. Before you agree to work with a broker, request a written disclosure of all terms and conditions for your dealings with the broker. Assuming that the broker is on the up and up, it doesn't hurt to explore this option.

If your development plan carries a risk factor that is higher than normal, a mortgage broker may be your best bet for financing. Many private investors work with brokers to lend money at interest rates somewhat above going bank rates. It might sting a bit to pay more for the money that you borrow, but if it is the only way that you can get funding, the increased cost may be justified. The key to working with brokers is to know, beyond any doubt, exactly what you are agreeing to before you agree to it. You would be wise to have your attorney inspect all paperwork before you sign any of it.

> **PRO POINTER**
> If your development plan carries a risk factor that is higher than normal, a mortgage broker may be your best bet for financing.

Private Investors

Private investors are a common source of money for developers. If you are working on a deal that will be difficult to obtain conventional financing for, private investors could be your ticket to success. Working with private investors may be more risky than dealing with banks. You can avoid many problems by having your attorney review all documents before you sign them. It is common for private investors to charge higher interest rates than banks. So long as you are willing to accept the rate of interest and the terms, there is nothing wrong with private deals. Know that some investors set very tight criteria that can hurt you if anything goes wrong during the term of your loan.

> **PRO POINTER**
> When you agree to terms for financing, you may not be thinking of potential problems. This could be a financially fatal mistake.

Assume that you have gone to a private investor and obtained financing for land acquisition and improvements. The deal that you enter into seems innocent enough when you sign it. You are happy to get the money that you want so badly. In your excitement you have failed to read all the fine print. As your project begins, you get involved in doing your job and don't think much about the financing terms. Your private lender provides money regularly in keeping with the terms of your agreement. Everything is going along fine. After several months of activity, your project is starting to take shape. You have done all of the hard work. Now it is a matter of putting the subcontractors on autopilot. What you don't know is that your world is about to explode.

You've been so busy taking care of business that you have been a bit remiss in keeping your files cleaned up. Paperwork has been piling up. You've been meaning to hire an assistant to help you stay on top of everything, but you just haven't had the time. As it stands, you missed two payments on your development loan. It's not that you couldn't have paid it; you just failed to stay on top of the administrative duties in your office. All your time as been spent on the job site. There is one piece of paper that you paid attention to. It was the certified mail that notified you of your loan default.

PRO POINTER

If you are interested in bringing in partners, builders are good sources to consider. Having you in the deal as the developer and the builders on board for construction makes a nice package.

You figured that the default was no big deal. You'd pay the payments and the late charges and everything would be fine. If you had been dealing with a commercial bank, your assessment of what would happen probably would have been correct. But it was not the case here, since you were dealing with a private investor. When you sent a check for the two payments and the late fees you thought all was well. But your payments were returned to you, and papers were delivered that announced that your property was in foreclosure. Whoa! Could this really happen? It could. Loan terms with private investors can be almost anything that all parties to the agreement agree to.

Builders and Contractors

Builders might be willing to work with you to get a new development up and running. They may lend you money or do work on speculation. If you are interested in bringing in partners, builders are good sources to consider. Having you in the deal as the developer and the builders on board for construction makes a nice package. Other contractors are also possible partners who may help either to finance your deal or to defer payment for their services until you start making sales. If you can get a site contractor to do the work either on spec or at a net-cost basis until sales come in, it can be a big financial help. It's customary to pay contractors who work this way more for the work that they do. You are increasing your expenses, but you are deferring the time in which the payment has to be made. Think of the increased cost as interest on a loan.

Some contractors are happy to buy into a development deal. They sometimes do it with cash and at other times with their services. Structuring deals with builders and contractors is a creative way to get a project going that would otherwise be stalled. As with all deals, get your agreements in writing and have your attorney approve everything before you offer it or sign it.

Partnerships

Partnerships are another way to raise money for a new deal. I've used partners before, but all my experiences with partnerships have ended with regrets. Personally, I would not enter into another partnership arrangement, but that's just my view. Using partners to launch a development can be a wise business move. Your partners might include any of the following:

- Your parents
- Your friends
- Business associates
- People involved in the development process
- General investors
- Your bank
- The existing landowner

Creative Terms

Creative terms can be set up for most development loans. Loans that are sold on the secondary mortgage market, like most loans for houses, must adhere to strict rules and regulations. However, loans that are not sold can be written to any variety of terms and conditions. Basically, whatever you and your lender agree to will work. This pertains to banks as well as private sources of financing.

> **PRO POINTER**
>
> If you go into business with partners, you will have to decide on what type of partnership to set up. You could go with a general partnership or a limited partnership. Most investors will prefer a limited partnership, since it limits their liability. Talk to your attorney to see how you should structure your partnerships if you decide to create them.

When you finance a development, there can be a lot of money at stake. How you structure the disbursement and repayment plans can have a lot to do with your success as a developer. Many first-time developers fail to think much about the terms of their loans. They are so consumed with the stress of getting a loan that they overlook key issues that will affect their projects. The way in which you negotiate the terms of your financing is crucial.

The schedule for disbursements is one of the first major considerations. You know the lump sum of your loan, but how will the money be parceled out to you? Will the lender advance you all the money you need to acquire the land? Will you have to put up a substantial amount of the land purchase with money out of your own pocket? A down payment of 30 percent is not unusual with land deals, but some lenders will advance the full amount of the land cost to get the ball rolling. This is certainly one of the first questions that you should be concerned about. Here are some questions you need to consider:

- How will the lender decide when to advance more money to you?
- Will the lender pay bills for you when work is done and invoices are submitted, or will you have to pay the bills and then request a draw against your loan?
- Will the lender pay in accordance with your costs, or will your bank draws be based on a percentage of completion?
- Are you going to be paying interest only when the total loan amount comes due upon completion?
- Will the lender allow interest to accrue with no payments being made for some period of time?
- How long will the lender allow your loan to remain active after the completion of your project?
- What are the terms and conditions dealing with late payments or missed payments?
- How long after missing a payment will you have to correct the deficiency before foreclosure proceedings begin?
- Which type of insurance policies will you be required to maintain to satisfy the lender?

Don't feel as though you have no say in how the terms of your financing will be structured. Some lenders may be very firm in their loan guidelines, but you should be able to negotiate some elements of your deal. It's well worth your time to think carefully about the terms and conditions that will work best for you. Some developers would rather pay higher loan expenses in exchange for less money out of pocket. Developers who have plenty of cash will prefer to get lower loan rates by putting more cash into a deal. Each developer has individual needs and desires. Establish what you want and then attempt to set up a loan that meets your approval.

> **PRO POINTER**
> You should insure your project to protect yourself, but your lender may require a more extensive policy that what you would be satisfied with.

CHAPTER SIX

Finding Suitable Development Property

When you start your search for suitable development property, you may find yourself overwhelmed with choices. Or you may feel that there is no suitable land available that you can afford. These two extremes are common. Many developers who don't have a lot of experience fall into the trap of wanting to buy everything that they see or not wanting to buy anything at all. If you don't purchase any land, your career as a developer will be difficult to pursue. Buying too much land or buying land under the wrong conditions can end your career quickly. Knowing what to buy, when to buy it, and how to buy it is very important to your overall success as a developer.

Some developers rely heavily on real-estate agents to find land. Brokers can be a big help to a developer, but leaving all property searches to brokers means missing some good deals. There are sellers who simply refuse to work with brokers. These sellers may have prime properties for sale. Developers who don't explore the world of private sales are sure to miss some good deals from time to time.

Developers sometimes hire private individuals to research available land. The researchers may be paid an hourly wage or based on performance. People who don't hold suitable real-estate licenses cannot broker deals between a buyer and seller. This doesn't mean that an unlicensed person can't do some digging to turn up prospective properties. If potential properties are located, the researcher can inform the developer of their existence. Then the developer can negotiate directly with the owner of the property.

Some landowners simply don't like to be bothered by real-estate agents and brokers. These same landowners might be very willing to talk to a developer or someone researching land for a developer. The use of a researcher is fine, but the extent of what the researcher can do without violating a state's real-estate laws varies. Ideally, the researcher should be a licensed broker who is representing the developer. This avoids a lot of potential legal issues. However, if the researchers confine their efforts to locating property and don't become involved in any way with negotiating a transaction, they don't need to be licensed real-estate professionals.

> **PRO POINTER**
>
> It is wise to work with real-estate brokers, but don't put all your potential success in their hands. As an independent developer, you have access to some purchases that brokers may never obtain listings for. And, remember, brokers are working for anyone who will buy what they are selling; they are not your exclusive land hunter.

Some developers prefer to do their own prospecting for potential properties. This approach has its advantages, but it takes a lot of time out of a developer's day. Running down leads on properties can be very time-consuming. Developers who do their own research have the advantage of knowing moment by moment how their searches are going. Effort is not wasted on property that a developer is not interested in, as could be the case if a researcher was being used. Most developers employ a multi-pronged approach to finding land. They look themselves when time allows, have arrangements with various brokers to always be on the lookout for land, and may have researchers doing some digging into private sales. Any of the methods can be effective.

Availability of Land

There may be an abundance of land available for possible development. Most of it will not be for sale or will not be suitable for a developer's needs. Ruling out land that is not right is often the first step in finding land. A lot of time can be spent chasing every parcel of land in sight. To conserve time and effort, some developers prefer to nail down exactly what they want and then look for it. Generally, having predefined parameters of what will be acceptable for development is more time-efficient and cost-effective.

Developers who wish to work in areas where development is already booming may be faced with a shortage of available land. When this is the case, developers may have to expand the radius in which they are willing to work or to alter their requirements for a piece of land. For example, a developer may have to look at what land is available and determine a way to make it profitable to develop rather than adhering to a preconceived development plan that will not fit the land available. Depending on circumstances, developers may have to alter their plans to fit whatever land they can buy. Doing this can be more risky than having a good plan that financial numbers support. Trying to mold a plan to a piece of land can lead to compromises that may result in less profit.

Where there is an abundance of land it can be tempting to try to control too much of it. A developer might try to tie up several parcels of land with options instead on concentrating closely on one piece of property. There are times when controlling a lot of surrounding land is a sound move, but it is usually safer to pick the best piece of land available and concentrate heavily on it.

Small developers should not diversify too much in their land purchases. Keeping all available funds available for a specific project makes more sense than spreading the money out among many potential projects. Entering into multiple projects at once is distracting for seasoned developers, and it can be destructive for new or small developers. Unless several pieces of property are being optioned as part of a single development deal, avoid the temptation to gobble up all the land in sight.

> **PRO POINTER**
> Before you start shopping for land, establish a detailed outline of exactly what you are looking for. You don't want to buy on impulse and regret the purchase later. Know what you want before you search for land.

Defining Your Needs

Defining your needs for development is an important step in the search for land. You could look for land and then decide what to do with it, but having a plan to follow during your search should prove more efficient and more effective. If you already have a rough idea of what you want to do, and you should, then refining the plan to better identify your land needs shouldn't be too difficult.

The needs for some projects are easier to define than others. For example, if you plan to build a facility where elderly people will be housed and cared for, you probably need a location that is convenient to medical facilities, stores, and the general population of an area. Building such a facility 20 miles from the nearest hospital would not be a good idea.

If you want to develop building lots to build log cabins from kit packages, city lots probably will not sell as well as rural land. The style of home indicates a need for more rural land. It would, of course, be possible to build log cabins in a city, but the people who would be interested in buying log cabins probably wouldn't want city lights and noises in their backyards.

Let's say that you want to create a development for professional office space. It is your intent to build a facility and lease space to medical professionals, such as dentists and optometrists. Do you want to look for land in the heart of the city? Would land on the fringes of the city serve your needs better? Is country land worth considering?

Country land is for the most part out of the question. Inner-city land is feasible, but will people want to fight city traffic or would they prefer a fringe location? Many developers have found that fringe locations sell and lease better than inner-city developments. To make a reasonable determination between inner-city land and fringe land, there is much to consider. Let's look at some of the considerations:

- What is the availability of land in each location?
- How do land prices compare between the two locations?
- Are there any zoning problems with either location?
- How populated are outlying areas?
- What does historical data show on the success of other fringe developments?
- Will city dwellers drive out to the fringe areas?
- What do demographics say about the income of area populations?
- Will you be able to provide adequate parking in both locations?

You could go on and on with the type of questioning that I've just described, and you should. Ask yourself questions daily. Don't assume that one brainstorming session is going to be enough to make a well-rounded decision. Think, think, and then think some more. Before you make an offer to buy land, you had better have a solid plan for what you are going to do with it.

Setting a Budget

Once you know the type of land that you want and where you want it to be located, you have to arrive at a maximum price that you can afford to pay for the land. Don't skip this step on the planning process. If you go out looking for land without an established budget, you may get caught up in an emotional rush that traps you into paying too much. Having a preset price limit will help you to stay within your profit zone.

Determining what you can afford to pay for a piece of land can be a complicated process. Many homebuilders try not to pay more than 20 percent of a completed project's total appraised value for land. I've have used the 20-percent rule-of-thumb for many years, and it has worked for me. But remember that builders are paying 20 percent for a buildable lot, not raw land that must be developed. As a developer, you must factor in all your costs to develop the land to a point where it can be built on.

> **PRO POINTER**
>
> As a rule of thumb, homebuilders generally try to keep their cost for building lots at a price that does not exceed 20 percent of a completed home's appraised value. This is important to developers. For example, if you are designing a subdivision for homes worth $300,000, builders will be unlikely to pay much more than $60,000 for a building lot.

The costs incurred to develop land can be varied and substantial. You have to estimate all the development costs of each individual project before you can arrive at a reasonable price to pay for raw land. These costs may include but are not limited to:

- Survey expenses
- Soils studies
- Loan-origination fees and interest expenses
- Engineering fees
- Permit costs
- The development and drawing of site plans
- Site work
- Legal fees

To expand on this, let's look at an example of a simple development deal. Assume that you are buying a one-acre tract of land that has a lot of road frontage.

Your plan is to cut the land in half to create two building lots. Since the land has road frontage along its entire width and there are no zoning complications, development costs should be low. After some research, you have found that basically all that you have to do is to have a survey crew divide the lot and draw a site plan for each lot that is created. There will be some legal fees and permit costs, but they will be minimal. Financing costs will be low on this deal. After pushing your pencil around for awhile, you estimate that all development costs will not exceed $5000. Now what do you need to know?

How much will the finished lots be worth? Can you estimate what the total appraised value for a house and lot will be? Assume that research shows a price range of between $150,000 and $165,000 per house. You want to be conservative, so you work with the lower number. Twenty percent of $150,000 is $30,000. You will have two lots for a total value of $60,000. If your development expenses are $5000, you will be left with $55,000.

How much profit do you need to make on the deal? Due to the type of deal that you are working with, very little effort or time is required of you. If you want a 20-percent profit on the gross sale price, you will be looking to make $12,000. Will you have to pay brokerage fees to sell the lots? Did you factor this expense into the development costs? Are you going to build the homes yourself? There are always more questions to ask and answer. For the sake of our example, let's assume that you anticipate paying a 10-percent real estate commission to sell both lots. This amounts to $6,000. You have $55,000 less $6,000, which gives you $49,000. You want $12,000 for yourself, so the number drops to $37,000. You could pay up to $37,000 for the raw land. But things could go wrong. Maybe you should factor in some percentage for mistakes.

Many developers would try to pay no more than $30,000 for the raw land. It is common for developers to attempt to keep their land-acquisition cost at half of what they hope the finished value will be. It's not always possible to maintain this type of spread, but it is desirable when feasible. Maybe you have so much confidence in the project that you will pay up to $35,000 for the raw land. The amount you are willing to pay is up to you. Deciding on values and profit percentages is a personal process. In some

> **PRO POINTER**
> It is common for developers to attempt to keep their land acquisition cost at half of what they hope the finished value will be. It's not always possible to maintain this type of spread, but it is desirable when feasible.

way, however, you need to peg a number as the most that you are willing to pay for raw land, and you should do this before you ever begin to look at land. Once you start shopping, you may get carried away with excitement and pay more for property than you should.

Working with Brokers

Working with brokers is easy if you can tell them what your buying requirements are. Since you are a buyer, you should be working with brokers who are buyer's brokers. This type of broker works for you but can be paid a commission by the seller or the seller's agent. If you deal with seller's brokers, you are not represented as well as you would be with a buyer's broker. Many brokerages specialize in buyer's brokers. Others specialize in being seller's agents, and some offer both types of representation. You should make arrangements to deal with a buyer's broker. Even with a buyer's broker you should also retain a good real-estate attorney to look after your interests in all your dealings.

If you retain a good buyer's broker, you should be able to leave most of the land searching to the broker. This doesn't mean that you should close your eyes to the market and wait for the broker to bring you a top-drawer deal. Doing your own research can pay big dividends. Depending upon how you structure your arrangement with the buyer's broker, you may not be obligated to compensate the broker for deals that you make directly with sellers when the seller's land is not listed with a brokerage. Don't make this assumption, though; have your full arrangement with the buyer's broker in writing and make sure that you understand the agreement completely.

Some investors and developers prefer to work with seller's agents in order to get more leads on land. When you commit to a buyer's broker, the broker does all the work with listed properties. If you prefer to have several brokers bringing you potential properties, it might be worth the tradeoff of not having full representation by a buyer's broker. If you decide to work with seller's brokers, you can send out a property description of what you are looking for to several brokerages and hope that some agent or broker will bring you a deal to consider. As long as you have good legal representation to review all your commitments before they are made, this procedure is fine. A really good buyer's broker will probably prove to be your best bet, but the shotgun approach with seller's agents will prove more effective if the buyer's broker with whom you are working is not aggressive in seeking out properties for your consideration.

The Do-It-Yourself Approach

The do-it-yourself approach to finding land is a proven technique among successful developers. You can contact numerous brokerages and make your desire known. If the brokerages have listings that fit your criteria, you will hear about them. You can search out land and contact owners directly to see if the land can be purchased. A trip to the tax assessor's office or the hall of records is all that is needed to obtain the name of property owners. Since you are doing all the prospecting yourself, you are assured of working towards the goal that you have set. Brokers sometimes try to slip in properties that are close to matching a developer's criteria without being a complete match. This is done in hopes of making a sale. You don't have to worry about this type of substitution when you do your own land search.

Time is one of the biggest drawbacks to the do-it-yourself approach. It can take a lot of time to chase down leads. You can go to a brokerage office and sit down with an agent to look through land listings. Most brokerages participate in some form of multiple listing service, which allows you to view the listings of all participating brokerages while going only to one brokerage. Newspaper advertisements can produce decent land deals. As more and more brokerages advertise on the Internet, you can surf your way through listings on the World Wide Web. If you have the time and determination, you can find solid land deals without the help of a broker. However, most developers find that working with brokers, in conjunction with personal searches, is the best way to find land quickly.

Appraisers

Real-estate appraisers are rarely thought of as professionals who are needed in a land deal prior to land being found. Developers who believe this are missing out on a great resource. Many appraisers are willing to work with developers in a variety of ways. The appraisers can help you determine what the finished value of a project will be. By looking at development plans and drawings, appraisers can get a good idea of what to expect the market value will

> **PRO POINTER**
>
> To assess property values, appraisers are the people to turn to for answers. Don't wait until you have bought land to find out what its market potential is; consult with appraisers in advance and know that your plans make economic sense.

be when a project is completed. Basic reports from appraisers can give you much insight into a project. Equally important is the fact that many appraisers will work as consultants to builders and developers. They are a resource that is often overlooked.

As a builder and developer, I have retained appraisers as consultants on many occasions. Their help has proved to be extremely valuable in my planning phases. Appraisers are in a perfect position to help a developer determine all sorts of information that may have an economic impact on a project. If you need to know the market values of homes in a certain neighborhood, an appraiser who works in the area can tell you quickly what you need to know. Are you curious about the vacancy rate of professional office buildings in the fringe areas of a city? The right appraisers can provide you with dependable numbers on vacancy rates. To assess property values, appraisers are the people to turn to for answers. Don't wait until you have bought land to find out what its market potential is; consult with appraisers in advance and know that your plans make economic sense.

Selecting Viable Projects

Selecting viable projects for development comes naturally to some developers. The process can be as simple as a gut feeling, but acting with so little to base a decision on is risky. Most investors and developers require extensive documentation on a property before a purchase offer is made. Once a developer begins searching for land, there is a good chance that he or she will have so many parcels to choose from that confusion will run rampant. If there are a number of real-estate brokers submitting properties for review, developers can fall even further behind in their decision-making roles.

> **PRO POINTER**
>
> Never jump on any deal just to get it before the next developer does. Some deals do require fast action to secure, but most of them will be waiting for you after you have done your homework on the property. It is better to miss a good deal than to buy into a bad deal.

There is no particular rule that dictates how a buying decision must be made. Some investors do operate almost entirely on gut feelings. Most investors have some type of system that they utilize to evaluate various properties. A few investors have strict guidelines that they will not stray from in the selection of real estate. Where do you fit in? Unless you have a very special gift of seeing into the future, you should come up with a system for ranking potential properties. Flying by the seat of your pants will probably cause you to crash and burn at some point. Proven systems tend to produce more consistent results.

Hard Sells

The first step to take in choosing viable projects is to resist the hard sells that some real-estate brokers will throw at you. Some agents have a flair for making a mundane property sound fantastic. With some salesmanship these agents can talk a lot of people into buying most anything. Be careful of this type of situation. You are the developer, and the decision about what to buy is yours. Don't let brokers or sellers try to convince you of something that you are not already sure of. Never jump on any deal just to get it before the next developer does. Some deals do require fast action to secure, but most of them will be waiting for you after you have done your homework on the property. It is better to miss a good deal than to buy into a bad deal.

Avoid deals that look too good to be true. Most opportunities that look like a steal have hooks in them. They are worth exploring, but don't make the mistake of entering into a purchase agreement without provisions for contingency clauses to protect yourself. Developers are good targets for people selling land. Since land can be difficult to sell, sellers love the idea of moving a lot of land in one quick deal with a developer. You have to protect yourself.

Developers are entrepreneurs and dreamers. They see visions of the future and get caught up in how they will change the world around them. These qualities set the stage for land sharks to feast on developers. If a salesperson can paint a pretty picture, some developer might just grab the deal without looking beyond the image that the developer wants to see. Sometimes the truth is much darker than the glorified vision of a land hustler or starry-eyed developer. The key is to avoid acting until you are convinced, based on proper research, that a deal is viable.

Competition

There is competition among land developers. They try to beat each other to the best parcels of land. To make your projects viable and less risky, you should avoid this type of competition. Compete against yourself. Force yourself to buy only into deals that will pan out well. If you are racing to beat a fellow developer to a deal, you are much more likely to make a mistake. The true competition between developers is settled when the projects are over and the outcome is weighed. If you have a great-looking project that was profitable, you are a winner.

Over the years I've seen developers buy land just to keep it out of the hands of their competitors. On rare occasions, this type action can be justified. For example, assume that I had a new subdivision nearly ready to build in. My plans called for building a complete subdivision of Victorian homes. If I found out that a competitor was about to buy adjoining land to develop for some purpose that would lower the value of my subdivision, I might be well justified in taking control of the property to keep the other developer from having an adverse affect on my project. Generally speaking, however, racing to a bank to buy land in order to keep competitors from getting it will result in losses rather than profits.

> **PRO POINTER**
> Developers who can see a new and better use for a piece of property have an ability to make more money than less visionary developers. Always look for the highest and best use. If you find it and the seller doesn't know about it, you have a chance to make substantially more money.

Average Deals

Average deals are offered for sale on a daily basis. These are deals that offer some possibility of profits, but the profit range is usually limited. Two ways exist to turn average deals into great deals. The first way is to buy land at a price below market value. This can be done, but it is seldom easy to accomplish. A better way, if you are creative enough, is to find a use for land being offered that others have not yet thought of. Basically, you are looking for the highest and best use of the land. If land is being offered as a large building lot, see if it can be subdivided into two building lots. When land is being offered for single-family use, see if local zoning laws will allow the construction of a duplex, triplex, or four-unit building. Turn every average deal around as many times as you can to see all of the facets of a land opportunity.

Average deals are abundant. They are rarely very profitable. If you have to pay full market value for land and then use it for the purpose for which it was sold, you will not make much money as a developer. The key is seeing below the surface and spotting opportunities that others have yet to see. Don't ignore average deals, but don't accept them on their face value. Find a way to make the land more valuable.

The Mass Market

The mass market offers a lot of land for sale. Pick up a newspaper and read the classified ads. Look in the little booklets available in convenience stores to see what land offerings exist. I have found some very nice properties in the weekly classified booklets that fill countless shelves in grocery and convenience stores. Check out the Internet; there are listings that cover nearly any need or desire. Spend some time driving around, looking for "For Sale" signs. Flip through a multiple-listing book with a real-estate agent. Everywhere you look there is land being offered for sale. The mass marketing of real estate opens the doors of opportunity to almost anyone interested in seeking rich rewards.

Are the mass offerings worth considering? Would you be better off searching out the unknown parcels of land that are not so obviously for sale? There are pieces of land that could classify as hidden treasure, and they are well worth finding. As for mass offerings, yes, they are well worth investigating. You will run through many listings that are not of interest, but you may stumble upon some that are right up your alley. For many investors and developers it only takes one good find to make all of the effort worthwhile.

Real estate that is advertised aggressively can receive a lot of attention. Potential buyers scour advertisements in search of the one property that they hope to find. The competition associated with mass offerings is substantial. Finding a prime listing among the many mundane listings can take months, if you find one at all. Time is spent looking at listing after listing. Is the time well spent? As soon as you find something that fits your needs, the wasted time on other properties will be forgotten.

Dealing with the maze of mass listings is easier if you have a formula for success. Developers who know what they are looking for and who can stay focused don't lose much time in skimming over the mass offerings. However, developers who are not committed to their game plan can waste a huge amount of time in weighing various offerings and considering a change in plans.

Much of what is offered for sale has some potential. If you spend all your time trying to shift with the land available, you may never get into the development stage. Being committed to a particular plan makes weeding through hundreds of offerings easy. Let me give you a quick example of what I'm talking about.

Assume that you want to develop a small subdivision of about 20 house lots. You want each home to have enough land to feel comfortable, and you need room for roads. After designing your development on paper, you have decided that you need a minimum of 18 acres of land to work with. Most of the land will be turned into house lots, but some of it will be consumed by roads, utilities, and possibly a community picnic area. You know that you want land outside the city limits but within a five-mile range of the city. Due to your plan, you want the building lots to be wooded sites. Access to the land must be from a good road that is near a major road leading into the city. You know that within the area of your search municipal water and sewer systems will be available. The criteria of your plan will hone your search results.

Given the example above, you can eliminate many of the land offerings that you may come into contact with through mass-marketing channels. As long as you stay committed to your buying criteria, sorting through the mounds of paperwork on properties for sale will not be a major task. As you skim through the pile, you can disregard properties as soon as some aspect falls outside your buying criteria. If you have a secretary or assistant, the screening process can be done for you. What you will be left with is a handful of potential properties that all fall within the limits that you have set.

By having a defined list of requirements for any parcel to meet, you can quickly eliminate most of the mass offerings. The more defined your buying criteria are, the easier it will be to rule out properties. Spending a little time planning what you are looking for can save you hours upon hours of time in the searching process. Once the basic stack of mass offerings is reduced to a manageable group of prospective properties, you can start taking the smaller stack apart, piece by piece.

Fine Tuning

Fine-tuning your selection can be fairly simple. If you are resolute about the requirements for a piece of property, it's simple to sort through properties. Unfortunately, finding property that meets every need you have can be very difficult. Most developers are willing to bend a little. If there are only minor discrepancies, you may be able to adjust your plans without restructuring your entire program to make a piece of land work for your needs. Developers who are willing to be very flexible have more trouble sorting through properties. Which group will you be in?

Sticking to your guns is a solid plan. As soon as you are willing to make exceptions, you expose yourself. But you also open more opportunities. You have to weigh the risk and decide how flexible you are willing to be. If you are going to stand firm on your land requirements, you don't need any help in learning how to weed out properties. When you find a property that meets all your criteria, you buy it. But if you are going to consider properties that don't fit your needs exactly, you may have a lot to consider, so let's talk more about this issue.

How far outside the parameters that you have set are you willing to go when considering properties? Can you be flexible on price? If you have decided that you want to buy property for half of what you expect to sell it for, you may be able to accept a property that is priced higher than you would like. The question becomes one of how much additional money you are willing to pay.

Money is not the only issue that you might be willing to negotiate on. Suppose you were looking for 18 acres and found a parcel that was perfect in every way except that it had only 16 acres. Can you live with less acreage? What would you do if you wanted wooded acreage and found a prime parcel that fit all your needs with the exception of trees? Any of these circumstances could be worked around for most projects.

If you are going to alter your plans, you have to factor in the changes and how they will affect your profit. For example, wooded lots would probably sell better and for a higher price than lots that don't have trees. A reduction in acreage is sure to affect your final income. How much a change affects your plans will depend on the type of change that you are dealing with. Before you agree to alter your plan, make sure what the end result is likely to be.

Going Around in Circles

Going around in circles is not productive. If you vacillate too much, the confusion will tear you up. Some developers change their minds so often that they might spend months accomplishing nothing. Viable properties are available. You have to know them when you see them. The best way to reduce the time you spend searching for properties is to define your needs closely and then stick to them. It can take a lot of self-discipline to adhere to your own rules. If you do, your odds of success are better, and you might be able to dig out some hidden treasure that is being overlooked by other developers.

CHAPTER SEVEN

What Makes a Project Viable

Planning on paper—or computer—is a sure way of making a project more likely to succeed. The benefits of a solid, written plan are numerous. Nonetheless, some developers don't have much patience for paperwork. They think of planning as boring and a waste of time. It is certainly not a waste of time, and it doesn't have to be boring. Many developers find the planning stage to be one of the most enjoyable aspects of land developing. There is some risk that you will find planning so fascinating that you will spend too much time on it. You have to know when to accept a plan as being as good as it needs to be. Some perfectionists keep tinkering with their plans for too long. It's possible that a developer who is unsure might use planning as an excuse to postpone the commencement of a project.

Many factors must be assessed as a plan is made for a development. Before a project goes into production, there will usually be a formal plan that has been created by experts. But developers need to create their own plans to give the experts a direction to move in. Not all developers do their own planning, and you don't have to. It's possible to hire people to do the planning for you. But doing it yourself can be quite rewarding. Even if you are not comfortable with preparing a complete plan on your own, you can work on many of the components within a plan. The more direction that you can offer your experts, the more likely it is that you will get a project plan that you like.

Components

The components of a master plan tend to be consistent from one plan to the next. However, different types of projects can call for different types of planning elements. Site features are a prime consideration. Both natural features and land constraints must be considered. Location may be one of the biggest factors to look at. In terms of location you might be concerned about how close a property is to schools, shopping, and medical care. You will also want to think about comparable properties in the area that you intend to develop.

After working out a plan that deals with existing features, you must create a plan for the development. Some examples to consider include:

- If you are building a residential subdivision, which house styles will you allow in the development?
- Are you going to have play areas for children throughout the development?
- Should you plan on paved bike paths?
- Will open areas provide recreational opportunities?
- Do you plan to build swimming pools and tennis courts?
- Will you be creating a high-density development?
- Would residents appreciate a community meeting hall within the development?

Good developers ask themselves hundreds of questions during the planning of a development.

People who don't have experience in land development generally have no idea how much effort is required to create a working plan. Anyone who has not developed land may not realize that someone has to think of everything from a name for the development to how the streets will be lighted. The amount of thought that goes into a development is tremendous. Minor details can have a major effect on a development, and failure to recognize a need can be extremely costly.

Demographics and psychographics both need to be considered when defining a neighborhood. The age, income, and family status of people can have a major influence on the type of development required. Developers must consider as many factors as possible to determine how to structure a development. For example, recreational needs vary with age. This could affect the need for walking paths or racquetball courts. Knowing your target market is the only way to build a development that can be sold at maximum profit.

Small Developments

Small developments have become popular. Residents often prefer a small development since it preserves a feeling of a close community. Developers sometimes feel that small developments are less expensive to develop. In terms of total cost, a small development does cost less than a large one. However, when the development cost is divided by the number of units available for sale, a larger development will often cost less per unit. This is due to the fixed cost of some expenses that remain essentially the same, regardless of the size of the development.

> **PRO POINTER**
> Tiny developments can fail since there are so few units to be sold. This is a risk with larger developments, too, but there may be safety in numbers. By this, I mean that if you have 20 building lots and 2 don't sell, you will probably survive. But if you have 4 lots and can't sell 2 of them, you could be in deep trouble.

Extremely small developments don't usually require the same types of expenses that larger developments do. Your budget may influence the size of your projects. But it can be just as easy, if not easier, to finance a large development as it is to finance a small one.

The appeal of small developments for buyers is their hometown feel. Large developments often seem cold and impersonal. If you are going to create small developments, you should plan on putting many personal touches into your planning. For example, the entry to the subdivision and the interior streets should be in harmony with the concept of the development.

Your Job

Your job as a developer requires you to come up with a plan that is unique. It should cry out for attention and offer plenty of appeal to your target market. The size and scale of a development must be planned. When establishing size and scale, you should consider whether they will be based on vehicular traffic or foot traffic. Development scale should be in proportion to the setting. Don't pick one set of criteria and use it throughout the community. It is better to vary and balance size and scale to keep a development more interesting.

Construction materials can have a lot to do with the tone of a development. What type of look are you hoping to achieve? Are you going to use a stone wall and pillars for the entrance to your development? Would a more modern look suit the nature of the development better? Many types of materials can be used to create different moods. Your development could be rustic or contemporary. It could be based on a high-tech motif or a classical setting. You and your experts have to decide on the image you wish to create and then choose materials that will allow you to accomplish your goal.

PRO POINTER

Architectural styles and land planning combine to define a neighborhood. You should strive to create a combination that will not become boring, but you should not diversify to a point where the development loses its overall tone.

Land planning is an area for experts, but you can start the process on your own. Assume that you are creating a development on the outskirts of a city. Your property is rural, rolling land that offers a country feeling, even though it is close enough to the city for a reasonable commute. The project is fairly large and will combine different types of living opportunities. There will be detached homes, condos, and duplexes. Based on the natural land conditions, the parcel looks perfect for a development of highly styled farmhouse designs. This design allows for large areas of living space and many options in exterior treatments. The building lots are large enough to accommodate big houses and ample parking.

Your development will be divided into sections for each type of housing to be built. Single-family homes will be in one area, condos in another, and duplexes in the third section. Since farmhouses are large by tradition, you will be able to build condos and duplexes that will give the appearance of a single-family home. Creative placement of garages and entry doors will make it difficult to tell one type of housing from another when driving through the subdivision. You can mix up the look of the development by using porches, varying roof designs, entryways, attached storage areas, garages, and so forth. All the housing will have the farmhouse theme, but there will be enough variation to keep the subdivision desirable.

As a continuation of the farmhouse theme, you might plan to use rustic appointments in common areas:

- If there is a stream for a footbridge to pass over, you might consider making a covered bridge.

- Bicycle racks might be built to look like hitching posts.
- Water fountains could be encased in small structures that resemble a covered well with a well bucket hanging from the rafters.

The creative possibilities are usually limited only by financial constraints.

You Make the Rules

As a developer, you make the rules that must be followed by people buying into your development. It is your decision on how covenants and restrictions will be used to maintain the look of your development. Your work should cover the following:

- Acceptable house designs
- Minimum square-footage requirements
- Allowable paint and stain colors
- Types and colors of roofing materials

Some developers go so far as to dictate the types of mailboxes residents may use. Without strong development guidelines, a development can lose its appeal quickly. You, the developer, are responsible for creating and protecting the development. Of course, you will probably turn to land planners and architects to assist you in the final stages of developing your guidelines.

Land Use

Guidelines for land use may begin with setback requirements. The setback refers to the distance that a building must be, at a minimum, from some other object, such as a property boundary. There are front, back, and side setbacks. For example, you may stipulate that houses must be at least 35 feet from any community sidewalk or street curb. You could say that no buildings may be erected closer than 15 feet to any side property line. And you might require a rear setback of 25 feet between a building and the back property line.

Another means of land control is to set a standard for how much of a building lot may be consumed by homes, garages, parking areas, and other impervious surfaces. Such rules can guarantee a certain minimum amount of green space around all homes. There are also regulations for site improvements, such as private sidewalks, patios, decks, and so on.

When you establish the rules that builders and homeowners must live by when in your development, you have to be very specific. For example, how wide can a private sidewalk be? Does your development have minimum and maximum widths on record? Wouldn't it look strange to have most sidewalks three feet wide and then to encounter some that were five feet wide? Will walkways have to be made of concrete, or will gravel be allowed? Consistency is important in development. Are covered entryways going to be allowed? Will they be required? You must decide.

PRO POINTER
Developers can dictate all sorts of rules for their developments. Too many restrictions can scare buyers away. But too few will make buyers nervous that a development may not maintain its standards. Even the smallest details, such as the type and style of house numbers used to display an address, should be considered when developing your master plan for a project.

Here are some other topics to consider when you are setting up rules and regulations:

- Swimming pools
- Storage buildings
- RV storage
- Lawn care
- Driveways
- Lighting
- Porches
- Decks
- Patios

Building Regulations

Building regulations are common for subdivisions. Depending upon the type of development, a developer may prohibit the construction of certain types of houses. Certain types of features may be prohibited. Some examples of prohibited buildings are as follows:

- You might prohibit the construction of single-level homes or homes built on slab foundations.
- Your rules might prevent builders from constructing houses that don't follow a specific theme.
- Rules could require certain roof pitches.
- There may be minimum square-footage requirements.
- Exterior trim styles might be specified.
- Paint and stain colors could be limited.
- Roofing materials might be required to be of a certain type.
- Window designs could be determined by a developer.
- A developer could require that all homes built in a development have brick foundations.
- The use of vinyl siding might be prohibited in a development.

Construction Regulations

Construction regulations must be planned in advance. Once a project is started, the construction requirements become a means of controlling workers and staying on time and on budget. Some builders and developers tend to overlook the construction rules. Don't fail to provide for construction procedures as you develop your master plan. What types of things do you need to be concerned about? Let's find out.

Temporary Construction

You may need to build temporary roads so that workers can enter the project. Don't omit this expense from your cost projections. Make arrangements early to allow workers access to your project. Getting a project ready to begin construction, then realizing that your crews can't get to the job site, is not only embarrassing but it sets you back on your schedule, and this cuts into your profits.

Temporary utilities are likely to be needed for your project. This will not always be the case, but determine if any utilities will be needed during the developing process. It can take several weeks or more to get temporary electrical service to a building site. If you think that this is a long time, just imagine how long it might take to provide services to a full development. You probably won't run into major needs for temporary utilities unless you are acting as the building contractor as well as the developer.

Depending on site location and conditions, you may have to install some type of retention system to control erosion. If construction equipment creates a lot of dust in a populated area, you may have to contract with someone to provide dust-control services, such as sprinkling construction roads. All these conditions must be accounted for in your master plan.

> **PRO POINTER**
> Fenced enclosures help to protect the expensive equipment used when developing land. Not all developers go to this expense, but you might want to consider the option.

Site Needs

There are certain site needs that may be required for your project. For example, portable toilet facilities are needed. Noise control is also a potential concern. Trees that are to be saved should be protected with temporary barriers. Some provisions should be made for the storage of equipment and materials. For material storage, many developers use trailers pulled by 18-wheelers. Construction office space is needed, so a site trailer should be placed on the property. Some developers buy their storage and office trailers, while others rent them.

Provisions must be made to control storm water and erosion during construction and development. A place should be established for the posting of required documents, such as permits and safety posters. Temporary signs will be needed, and you may need someplace to post them. Sit down and figure out all that you can about your needs.

> **PRO POINTER**
> The management of a project is crucial to the success of a development deal. Someone must make sure that all permits are obtained and posted. Keeping active insurance on a project is essential. Any subcontractors used for a project must be screened for insurance and general business compliance. Time schedules must be set. Work hours for contractors must be established. The management needs of a project can be extensive.

Project Manager

You must decide if you will serve as your own project manager or if you will hire one. Good project managers don't work for peanuts. If you are going to hire one, you must factor the overhead cost into your overall budget. Maybe you are not sure if you can do the work yourself. If you have any doubts about your abilities, factor in the cost of a project manager. It's better to have the overhead factored into your budget and not need it than it is to have the expense missing and then find that you need a manager.

The planning stages of a development can be tedious. If you prefer to have it done for you by experts, go that route. Most developers do preliminary planning on their own, and it is good experience. Once you have a general plan, you can turn it over to the experts. Then, as results come in from the experts, you can compare the final plan with your rough draft. Look to see how much has changed. When you get to the level of minor or extremely technical changes, you will know that you have a good handle on what it takes to be a successful developer.

CHAPTER EIGHT

Looking For Land in All the Right Places

Is finding hidden treasure in real estate still possible? Absolutely! There is plenty of land waiting for some aspiring developer to grab it and turn it into powerful profits. Times have changed over the years. I remember riding around with my grandfather when I was a young child. My grandfather was retired and enjoyed working with real estate. He was not a big-league developer, but he did plenty of deals, many of which he made from either the front seat of his car or in a rocking chair on a porch. Riding around from farm to farm with my grandfather was my introduction to real-estate dealing. As it turned out, it was an education that I still use and that has done well for me financially.

Over 35 years ago, when I was going door to door with my grandfather, I learned the value of prospecting. My grandfather's name was Paul, but I called him Amos. He was a very special man in my life, and the lessons I learned from him close to four decades ago still pay off today. Back then, the game was to carry a large roll of cash around to flash on farmers who might be willing to sell off part of their land. When Amos would see a piece of land he liked, he would drive up to houses and start talking to people. Sometimes they owned the land, sometimes they didn't. But they almost always knew who owned the land. This was the first step in mining the gold from raw land.

I can still see the old farmers in their bib overalls. Sometimes they were working their fields, feeding their cattle, or rocking on their porches. We would pull into their driveways and blow the horn on the old Chevy. Dogs would come out barking, and then someone would amble over to the car. Occasionally we would be invited to the porch for lemonade or iced tea. More often than not, negotiations started right at the car window. After basic greetings Amos would get right to the point and start talking business. If the farmer showed any interest in selling some land, Amos would whip out his roll of cash and flash it. Hard cash always got the attention of the landowners. The visual impact of greenbacks was much stronger than a cashier's check would have been, even if it had been in a larger amount.

My grandfather's buying strategies were sound. He would ride around and find property that he felt had potential. Then he would canvas an area to find the owner and ask if the land could be purchased. There were plenty of rejections, but there were a lot of deals made, too. The land that Amos went after was never listed with brokers or advertised in newspapers. Finding land that was not known to be for sale made it easier to get a good deal--there was no competition. This type of approach still works today.

Dealing Directly

Dealing directly with landowners is one of the most efficient ways to acquire quality properties if you can locate the land and the owners without too much effort. Sellers will often sell for less if they don't have to pay a brokerage commission out of their sale proceeds. Sellers who don't have to pay commissions or advertising costs can sell for less and still net the same amount of income. Some sellers are extremely willing to work directly with buyers but not all: If you plan to search out properties on your own and to go directly to landowners, you have to be prepared for a lot of rejection.

People who make their livings selling things know very well what cold calling is. It is one of the worst parts of selling for most salespeople who have to do it. Cold calling is a numbers game. I remember one company that I used to work with that tracked the results of cold calling. Their formula was that it took 10 calls to get one appointment to meet with a prospect. Company requirements were to get eight appointments a day. This could mean making 80 or more cold calls daily. It was no fun. Going door to door to buy land is not much different from calling people and asking them to change their long-distance phone service. Many people hang up on cold callers, and some landowners are not much more amiable to prospecting developers. I don't tell you this to turn you off the idea of knocking on doors, but you have to know that you will meet with a lot of rejection as a roaming prospector.

Some developers get lucky and make a deal almost as soon as they begin talking to landowners. This is the exception rather than the rule. If you are not comfortable with calling people whom you don't know or knocking on their doors, you can use a letter to introduce yourself. I prefer direct mail to cold calling and door knocking. Letters are not as effective, but they are faster and, if you have enough land to go after, the odds are in your favor.

> **PRO POINTER**
> Direct mail can be a powerful prospecting tool when looking for land to buy. Find the region you want to develop and mail every landowner in that region a letter of interest in buying land. You might be surprised by the results.

I moved from Virginia to Maine in 1987. Maine is a state that sees a lot of vacationers. It is common for Maine land to be owned by people who live outside the state. As both a broker and an investor I have used this to my advantage. I have gone to the tax assessor's office and found the names and addresses of all landowners within a tax district. A quick skim of the list shows which owners live out of state. These people become my first target. A well-written letter mailed to out-of-state owners can result in some fast action.

I've done mass mailings to out-of-state owners and received responses in less than a week. Since the owners are not in Maine, they sometimes tire of owning property and paying taxes on it when they can't use or enjoy it. This makes the sellers eager to entertain offers. In many cases the tax offices have supplied me with pressure-sensitive labels with the mailing addresses of landowners. This makes a mailing campaign easy. I have used direct mail successfully for many years. You can too.

Ownership records of real estate are public knowledge. This means that you have access to the information. The local tax assessor's office is the first place to start when seeking the name of an owner. If you have a property address, getting the name and address of the owner is simple: You just look it up in the tax records. If you want more information, you can take the data you get from the tax office and research the deed on the property. Nearly all real-estate deeds are recorded. Once the deeds are recorded, they become open to the public. Deeds will tell you how large a parcel of land is, what easements exist, what liens are filed against the property, any covenants and restrictions that may exist, and so forth.

Deeds are usually kept in local courthouses. Some regions keep their deeds in what is called a hall of records. A phone call to the local tax office will put you on the right track to find where deeds are kept. Normally, deeds are kept at the county seat, not in local town offices. Check with your local authorities to find out where you should go to look for deeds. Whether you knock on doors, make telephones ring, or do your prospecting by mail, private owners can be a great source of outstanding land deals.

Back Taxes

Back taxes can be a serious problem for some landowners. Since raw land doesn't usually produce income, it can be hard for owners to justify keeping the property. Sometimes owners become delinquent in their tax bills. If tax bills go unpaid for long enough, the taxing jurisdiction may take the land and sell it to collect the unpaid taxes. Buying land from tax sales is one way of getting low prices, but finding owners who are in trouble with taxes and making deals with them before a tax auction may be a better way to cash in on the opportunity.

> **PRO POINTER**
>
> Property owners who are having trouble paying their property taxes may be very willing to work with you. It may be possible to buy land from these sellers at below-market rates and with attractive owner financing.

If you find and contact people who are behind on their land taxes, they may be happy to sell their land to you. In some cases you might be able to buy the land with excellent owner financing. For instance, you might pay the back taxes as your down payment and then make periodic interest-only payments on the land until you are ready to begin the developing process. Reviewing records in the tax assessor's office should reveal properties that have outstanding tax bills due.

Auctions

I have never found auctions to be a good source of land sales for development. If you are looking for houses, apartment buildings, commercial space, or even farms, auctions can produce good results. However, finding the right land for development at an auction is a crapshoot. Another problem with properties sold at auction is that payment is expected for the full purchase price in a short period of time. This usually is not compatible with the needs of a developer. Some real-estate investors do very well by working auctions, so you may want to check this option out. My personal experience has not shown this method to be effective enough to concentrate on, but you may encounter different results.

Foreclosures

Foreclosures, like auctions, have not proved fruitful for me as a developer. Buying property that is in foreclosure is a good way to get a low price, but finding property that you want to develop that just happens to be in foreclosure is difficult and time-consuming. Like tax auctions, waiting for a foreclosure auction is not the best way to deal. It is often possible to become known by lenders as an investor who is always interested in foreclosure properties. Getting to know the people at lending institutions who are responsible for liquidating foreclosed properties could pay off big for you.

Any lender who loans money on land is sure to handle some foreclosures. Generally, there is a special officer within a lending institution who is in charge of properties that are in default of loan payments. Sometimes these people have the title of work-out officers. The term comes from the job description of working something out with someone to protect the lender's financial interest in real estate. You could call lenders and ask to speak to the person in charge of foreclosures. This procedure will work, but there may be a better way.

My approach to foreclosures has been to go in a side door. Normally, I meet with loan officers first. I explain to the loan officers what I do and investigate the opportunities of borrowing money for my ventures. As these talks progress, I ask the loan officers about how foreclosures are handled at their company. Once a relationship is formed with the loan officers, getting to the work-out officers is easy. A side benefit of my approach is that a path is already made for financing when I find a foreclosure property that I'm interested in.

You may find that the effort required to turn up foreclosures is not rewarded often enough with prime development properties. This has been my experience. But you might walk into a bank on just the right day and find a diamond in the rough. Also, depending on the type of developing that you are doing, you might be interested in dilapidated buildings. If your development plans call for small lots, you might find rundown, abandoned buildings that can be torn down to give you a new use of the land. When this is your angle, foreclosures and auctions can both be very productive. Each developer must create an individual

> **PRO POINTER**
>
> Talk to banks and get to know the people who handle delinquent accounts that are going into foreclosure. This can be an excellent way to work out great deals both in terms of price and financing for the purchase of land.

system that works on a personal basis, so consider every option once you define your development plans.

Lawyers

Lawyers can be excellent sources of leads for land. When estates have to be settled, lawyers are usually involved in the process. If you become known as someone to call when there is land to be sold, you may have your phone ringing regularly with lawyers who have land to sell for their clients. Seasoned investors work with lawyers frequently, and once you are in the loop, a lot of opportunities can come your way. Properties bought in this manner are sometimes less expensive due to the savings on marketing and brokerage fees.

Some investors who are not familiar with working with lawyers concentrate on real-estate attorneys. However, estate attorneys are the ones to go after. Use real-estate lawyers to represent you when buying and selling real estate, but look to estate lawyers for good land deals. A simple letter of introduction mailed to all the lawyers in your area can be enough to get the ball rolling with estate sales.

Accountants

Accountants are another source of potential hidden treasures in real estate. Landowners frequently consult their accountants for advice on the tax advantages and benefits of selling real estate. If you are listed with a number of accountants as a potential source for buying real estate, you can be one of the first people called when a landowner is thinking of selling. Not all accountants or attorneys are willing to give their clients the names of potential buyers, but many will. The process is not like insider trading or some other form of questionable conduct. There is nothing wrong with providing a list of potential buyers to clients who have real estate to sell. If a professional, such as an accountant, was giving out only your name, that might raise some ethical issues, but it doesn't hurt to get your name on the list of potential buyers for professionals who do offer such information to their clients.

Benefits

There are many benefits of getting to sellers before they list with brokerages and before they begin advertising their properties for sale. If property is not on the open market, the level of competition among buyers is much lower. Sellers who don't have to pay real-estate commissions or marketing expenses can sell for less and still be happy. You will work hard if you seek out private sales, but the benefits may be good enough to warrant the extra time and effort. Once you find property that you are interested in, getting control of it is the next step.

CHAPTER NINE

Initial Investigations

Initial investigations are a needed part of any successful land-development plan. Before you buy property, you should involve a variety of experts in a property evaluation. These people are likely to be soils engineers, surveyors, land planners, and related professionals. Professional fees for these experts are expensive. Before you burn up substantial cash in professional fees, do some initial investigations on your own.

Many specific topics of potential concern are covered in other chapters of this book. Here we will consider how to make sure that a project has the proper potential to be a sustainable development. When you begin investigating prospective properties, there are questions that you should ask yourself. With this in mind, let's start with some of the questions that you should consider.

Space Preservation

Space preservation is always a consideration in any type of development project. Will the land that you are looking at support a compact community that will be surrounded by a sustainable environment? Will there be adequate open space remaining after development? To get answers to these questions, you must do some thinking and planning. Start by making a site inspection.

Visit your proposed land venture with a camera and a tape recorder. By taking numerous photos and recording your thoughts on the site, you will be able to make a valuable review of the land once you return to your office.

Walk the boundaries of the property and make notes of all topographical details. Look for streams, hills, trees, and other factors that will come into play as you develop the project. As you walk the land, envision what you will leave as natural space. How much land will this leave for building? Compact communities are growing in demand. You and your professionals will have to create a viable plan to match your natural-space requirements with living space.

Will You Combine Mixed-Use Properties?

Will you combine mixed-use properties in your sustainable development? A lot of developers look at land development through tunnel vision. They see only residences. This can be a big mistake. Communities need services. You should consider developing a section of land to house convenience stores, grocery stores, gas stations, and other types of necessary commercial space. By providing these stores, you allow your residents to reduce traveling and in essence minimize the impact on the environment from driving greater distances.

Urban Infill and Brownfield Redevelopment

You can add to the value of your development by including urban infill and brownfield redevelopment. This is an excellent way to reclaim ignored property. The cost can be extensive, but you may be able to obtain financial aid for the process that will be beneficial to you, while the reclamation of the land is good for the community. Check with your land experts to see if this is a viable option for your project.

Preplanned Growth

A sometimes overlooked element of sustainable building is preplanned growth. If you crowd the development and limit natural space, you will have a negative impact on your concept. It is always tempting to try to squeeze just a little more out of your land. Keep in mind that the reason for success in the first place may well be the result of expansive green space. If you dip into this space to build more structures, you may ruin the appeal of the development.

Erosion Factors

Are you looking for erosion factors in the parcel of land that you are contemplating? Erosion and sedimentation during construction must be taken into consideration. When going green, you want to look for natural means of avoiding erosion. This generally has to do with topographical characteristics that are already in existence. Yes, you can have site work done to build the equivalent of natural barriers, but this will add to your development cost. Try to find a site that will not require substantial site work.

Irrigation

If you are developing in a region where irrigation of lawns and grounds is common, you should consider the natural resources of the land. Is the lay of the land suitable for storing reclaimed water for irrigation? How difficult will it be to create irrigation ponds? Will the water reserves be far enough away from structures to avoid potential liability risks? There is a lot to consider when you are conduction initial investigations.

Summing Up

After your site visit, start summing up your findings. The work that you do now can save you money when you call in the professionals that you will need to make final decisions. You may not have the experience to make a decision to go forward, but if you see enough negative factors in your first investigation, you can eliminate the subject site and continue your search.

Compile your notes and put them in writing. Keep your notes organized. Lay out your photos. Pore over the material and then walk away from it. Come back to the material and evaluate it a second time. A third look is also a good idea. Check your data more than once, but don't look it over so often that it all becomes a blur. Once you have a strong overview in mind, put your thoughts on paper. Keep your work neat. In the end, you will be turning your photos and notes over to various experts for their professional opinions.

If you decide after considerable thought to move forward on the acquisition of a parcel of land, the next step is to get your notes and photos into the hands of your experts. Your preliminary work will save you a lot of money if you rule out a property, and it will save you some money if you move forward, as you will have given your experts a head start.

Planning is of key importance in successful land development. Don't overlook the value of thinking. You can have a system and you should, but the system is only as solid as the person implementing it. With this said, let's move to the next chapter and consider drainage factors.

CHAPTER TEN

Drainage Factors

Dealing with storm-water drainage is important when you develop a piece of land. As you alter the land, you create new needs for handling runoff water. The size and nature of a development has a direct impact on how sophisticated a storm-water plan must be. In the case of a single house, the creation of a drainage system is usually easy. Most builders surround their foundations with slotted drainpipe that is bedded in and covered by a layer of crushed stone, which is further covered with dirt. The drainpipe might run to a sump location for pumping, but it will generally be piped to a suitable discharge location with a gravity installation.

Gutters installed on houses and piped to underground drains are very effective at keeping rainwater from eroding the soil around a foundation. The subsurface drains for gutters may tie into the foundation drains that are carrying storm water to its discharge pump. If a sump pump is installed in a house, the discharge hose from the sump pump may also intersect with the underground drainage piping at some point. The discharge location for the drainage pipe may be a dry well or a storm sewer. Check local code requirements to see what options are available to you in this regard. In general, the handling of storm water for a house that sits on a well-drained lot is easy and not very expensive. However, most larger projects require somewhat more effort, thought, and money to protect from excessive storm water.

While a single residential project can be protected from storm water with a simple, inexpensive system, a major development can require extensive work. The engineering installations for major projects can be quite expensive. There are multiple factors that must be considered in a storm-water-management design. Here are the four primary concerns to be considered during the planning phase:

- Peak flow changes
- Quality of lakes and rivers
- Total runoff water
- Quality of groundwater

> **PRO POINTER**
> It's not enough to contain or remove storm water. The process must include a plan to protect other types of water.

Building a development can create many changes in the natural dispersion of storm water. Think about it. If you install paved parking areas, homes, sidewalks, and other impervious barriers in a natural area, you are bound to create more runoff water. There simply isn't as much earth available to absorb the water.

When engineers are developing a plan to deal with storm water, they must consider the effects that their plans will have on all other types of water and related concerns, such as:

- Drinking water
- Groundwater
- Lakes
- Streams
- Ponds
- Sediment
- Erosion control

Every parcel of land has its own natural flow routes for storm water. Some of these routes may be destroyed during the development process. If you can maintain the routes during and after developing, you might reduce your overall development cost. For example, if there is a natural wet-weather stream that runs quietly behind your major development area, you might be able to use the natural element of the land to funnel water to a retention or detention pond. You might run into environmental problems, though, so have your environmental experts confirm that your drainage plan is acceptable to all governing authorities.

With the right designs, storm water can be recycled. Building ponds to catch storm water and then using the water to irrigate the land or a golf course is a great idea. Most developers prefer using multiple ponds rather than a single pond. The proper use of landscaping helps prevent erosion and can slow the rate of flow for the runoff. There are many techniques that can be used to make storm-water systems both attractive and effective. Your engineers will have to decide between a closed and an open system. Most people agree that open systems are more desirable. Let's discusses the differences between the two systems and see what you think.

Closed Drainage Systems

Closed drainage systems are generally considered less desirable than open systems. Cost and maintenance are two factors that come into play when choosing a system type. The installation of a closed system usually requires extensive piping. Closed systems are designed to be run underground and out of sight. The components of a closed system can include pipes, catch basins, inlets, and both retention and detention areas. Some developers like closed systems since they are not visible. Keeping the drainage system out of sight is preferable at times.

> **PRO POINTER**
> Most major drainage systems installed below ground require catch basins. These basins are expensive and should be kept to a minimum.

If you opt for a closed drainage system, you have many factors to consider. Most developers start by planning the routes of drainage piping. You or your engineers must look at the area to be drained. Next, a decision will have to be made for routing and sizing pipes. Planning the details of a storm-water system is beyond the capabilities of most developers. Some master plumbers possess the ability to create a system, but the design work is normally done by engineers.

Sizing pipes for a drainage system can be a complex task. The formulas used are normally provided in the local plumbing code. Total expected rainfall must be calculated, as well as maximum runoff potential. Piping must be sized to accommodate maximum water flow. In the case of site drainage, inverts are needed to direct surface runoff into the underground piping. Inverts are merely devices that collect surface water and deliver it into a drainage system. Grates are used to cover the openings of inverts. Most local jurisdictions require the use of sediment traps in conjunction with inverts. Sometimes filters are used in place of traps. State and local codes dictate the precise requirements for sediment control. Developers must have a good idea of what will be required for drainage before a reasonable cost estimate for the infrastructure of a project can be completed.

Most major drainage systems installed below ground require catch basins. These basins are expensive and should be kept to a minimum. Proper design can reduce the number of catch basins needed for a closed system. The basins are meant to retain sediment as storm water passes through a drainage system. Catch basins may be called for under inlets of a system, at changes of direction in a piping system, and close to streets.

> **PRO POINTER**
>
> It is best to route piping to avoid placing it under buildings and other elements that would make access to the pipe difficult. Since closed systems are installed below grade, they can be expensive to repair. Don't add to this expense by making access to the pipe difficult to obtain.

Most catch basins are made of formed concrete. The concept behind a catch basin is similar to that of a septic tank. A pipe delivers water to the basin and allows the water to flow into the holding tank. Heavy sediment sinks to the bottom of the enclosure. An outlet pipe is installed at some distance above the bottom of the tank but below the inlet pipe. As sediment drops to the bottom of the catch basin, water is allowed to flow into the outlet pipe and on through the drainage system.

Piping for storm water should be designed to minimize turns. As with any type of plumbing, bends in the piping can cause stoppages in the flow of water if debris enters the system.

Cost is a prohibitive factor for closed systems. Open systems are generally less expensive to construct. However, when the water flow is too great for an open system, a closed system must be used. In cases where an open system is considered to be an eyesore, a closed system is the answer.

Open Systems

Open systems for the control of storm water are common. Open swales carry water to retention or detention areas. These systems work well for some projects. If the maximum amount of runoff is not excessive, an open system is a cost-effective option. The overall design of a development can have a lot to do with the viability of an open drainage system. Most developers strive to make their open drainage systems as attractive as possible. This can mean added expense, but the end result is often worth the cost.

Drainage Factors

It would be easy to run a swale in a straight line to a retention pond. Getting from point A to point B with a straight line is both the most direct route and the least expensive path. But it probably is not the best path. It is generally better to create multiple ponds rather than one large pond. Winding swale paths also tend to look more natural, and they provide more opportunity for vegetation and attractive design.

When ponds are created, the edges should slope gently to the center. Steep drop-offs increase the risk of drowning. Ponds that deepen with gentle slopes are less threatening. Landscaping in and around the pond is also important. When a detention pond is used, there will be times when the pond is not holding substantial water. Without the proper landscaping, the bottom of the pond will be an ugly eyesore of sediment and mud. Planting the bottom of the pond with vegetation that will thrive in and out of water makes a low-water condition more attractive. The vegetation also acts as a filter when water enters a pond.

Given a choice between open and closed systems, open systems will normally be less expensive and easier to maintain. In addition to lower costs, open systems can provide relaxing recreational opportunities. There are times when a combination system is required for proper site drainage. Each site is different. You and your engineers must spend enough time during the design process to create the best solution for drainage that is available to you.

The deciding point on the type of system to use is often determined by the amount of runoff water expected. High volumes of runoff call for a closed system. Residential developments normally leave more land in a natural state than commercial sites. It's more common for runoff to be higher on commercial sites. This indicates a likelihood that closed systems will be used with commercial projects, while open systems may be used with residential projects.

Let's assume that you are working on a residential development. The project is large and will contain a variety of amenities. Many of the housing areas

> **PRO POINTER**
>
> When designed properly, detention and retention ponds can become an enhancement to a development. Retention ponds can be stocked with fish to give residents a source of recreation. In some cases retention ponds can be used as ice-skating areas in winter. Size and depth of ponds vary with the drainage needs of a project. Having deep ponds in residential areas can pose a hazard. Residents may shy away from the development due to a fear of drowning.

will have limited runoff and can be served by an open drainage system. Some parts of the development will have substantial pavement and concrete. These areas will produce rapid, high-volume runoff. This type of project will need a combination system. You will use a closed system for the paved areas and an open system for the natural areas. Of course, your engineers will make final determinations on what your project will require.

Looking at the Land

Looking at the land that you plan to develop can tell you much about drainage needs. A visual inspection can reveal runoff patterns. Of course, your development alterations will change current trends in the runoff water. Amazingly, some developers fail to think of this. The land that you are looking at in its raw form is not the project that you will be draining. You and your experts must consider what the project needs will be once development is complete.

Your engineers can come up with a good preliminary plan for drainage early in the planning stages. The engineers will have to know the master plan, street locations, types of buildings and their placement, and other factors that affect drainage. Topographical maps are used to trace natural water flow. Calculations are made based on existing conditions and proposed changes. It is during this stage of planning that engineers may be able to make adjustments to lower the cost of drainage for storm water. Minor changes in construction plans could save you considerable money in the infrastructure of your development.

Sectional Designs

Sectional designs are not uncommon for developments with mixed landscaping. Obviously, a project that has varying topography and different levels of development will need different types of drainage control. You might break your development down into sections and then design a drainage system for each. Part of the design may be a closed system, while the rest is an open design. You might use one large retention pond to control drainage in one part of the development, while a series of smaller detention ponds might be used in another section.

The amount of development in an area certainly affects drainage. A lot of concrete or pavement creates different needs than a wooded section of land. Various soil types also affect the needs for runoff drainage. You will need soil studies to determine how much runoff water a section of a development can handle under given conditions. Whenever soil types differ, you must alter your plans for drainage control.

Engineers must factor in many considerations when developing a drainage plan. As a developer, you should rely on your experts to provide you with a suitable storm-water system. Be aware of the basics so that you can talk with your engineers and understand most of what they present to you. Don't attempt to design your own drainage system when dealing with a sizable project. Your job may not require extensive drainage. Maybe you will need nothing more than a driveway culvert and some subsoil drains for gutters. If this is the case, consider yourself lucky. However, if you are doing a major project, the engineering and installation of a storm-drainage system will be an expense that you can't afford to overlook.

> **PRO POINTER**
> You will need soil studies to determine how much runoff water a section of a development can handle under given conditions. Whenever soil types differ, you must alter your plans for drainage control.

CHAPTER ELEVEN

Soil Considerations

Soil considerations are best left to soil scientists and engineers. But it never hurts to have a good idea of what the experts will be talking about. On some sites the primary concern is that the soils perk well enough to allow the installation of a septic system. Large sites can be much more involved. In many cases the soils control much of the development potential—not to mention the cost of a project. Many factors influence the viability of a project, and the soil conditions are certainly a major consideration.

Most builders and developers are not overly concerned with scientific data pertaining to soils. If the ground will support foundations, drain properly, and support a successful development, that is generally all that builders and developers are interested in. How much do you want to know about soils? Do you really care that the official name for a soil scientist is an agronomist? Probably not. Even if you don't care to know the differences among soil types and characteristics, you should invest some time in understanding the basic terms that you will be dealing with as a developer.

What is soil? Do you know what soil is made up of? There are three components to soil. The first element of soil is the particles that most people think of as dirt. But mixed in with the particles are water and gases. Soil that is fully saturated has all the voids between the particles filled with water. If the soil dries out completely, the voids are filled only with gases. Under normal conditions both water and gases fill the voids between the particles. Engineers must assess the phases of soil to arrive at their data. The relationships among gases, water, and particles must be established in a weight-to-volume ratio. Once engineers have the basic evaluation of the soil, they can establish the soil properties for such elements as shear strength, shrinking, swelling, consolidation, and so forth.

Texture of soil is an important characteristic. Many people refer to soil based on its texture. How many times have you heard people talk about sandy soil or clayey soil? The texture of soil is determined by the relative amounts of the individual particles making up the soil. Different tools are used to identify particle sizes. Sieves are used to sort particles by size. Hydrometers are used to measure the amount of soil in suspension. The process is much more complicated than most builders or developers are willing to become involved in. That's why we have engineers! Without getting into scientific formulas and procedures beyond the duties normally associated with land development, let's look at how the soils on a project will affect your business.

> **PRO POINTER**
>
> Both alluvial and lacustrine soils tend to make poor foundation materials. They are soils generally composed of medium to fine sand, silt, or clay.

Types of Soils

There are many types of soils to be encountered. Many soils are transported in one way or another. The physical aspects of soil types can be judged to some extent by the means of transportation. We're not talking about trucking dirt into a site. No, the transportation referred to here has to do with natural movement.

Alluvial and lacustrine soils are created by sedimentation. Alluvial soils are left by running water. Lacustrine soil is the result of deposits in lakes. Since alluvial deposits are transported by running water, a natural filtering process takes place. Large particles tend to sink to the bottom of a stream, while smaller particles are moved with the water. The separation process is natural and effective. Both alluvial and lacustrine soils tend to make poor foundation materials. They are generally composed of medium to fine sand, silt, or clay. The drainage factor for this type of soil is poor. Building a foundation on alluvial or lacustrine soil is risky, since the soils are usually soft, loose, and highly compressible.

Glacial soils are called moraines. The soils are pushed, eroded, or carried along with glacier movement. Particle sizes range from a clay consistency to large rocks. Characteristics of glacial soils can vary greatly. However, glacial deposits generally offer a good base for foundation construction.

Aeolian soils are transported by wind. As you might imagine, this means that the soil must be small and light. Sand is the most common type of aeolian soil. If you have ever been to a desert or a large beach, you have probably seen sand dunes, which are representative of aeolian soils. Sand is not a great foundation soil to work with.

> **PRO POINTER**
>
> The bearing capacity of soil is a primary concern when planning a development. You must know what the soil's ability is to support structural loads created during construction, such as buildings, roads, and other improvements.

Colluvial soils are transported by gravity. In most cases this type of soil is the result of hillside deterioration. Rock chips comprise most of a colluvial soil. Since the soil has moved once, it is likely to move again. This makes it undesirable as a foundation material. Organic soil is no better. This is soil that is made up of decaying plant life. Peat is the best-known organic soil. Due to its compressible nature, it is unsuitable for foundation construction.

Bearing Capacity

The bearing capacity of soil is a primary concern when planning a development. You must know what the soil's ability is to support structural loads created during construction, including buildings, roads, and other improvements. A first concern is the strength of existing soil, but you will also want to know how excavated soil that might be reused on your project will hold up. When embankments will be created, you have to know how stable the soil used to build the embankment will be. If you find that some of the soil is substandard in strength, your engineers might be able to recommend a way of improving its bearing capacity by adding chemicals or materials to the soil. In any event, you must establish that the soil will be suitable for your development plans.

Bearing capacity is increased when soil is compacted. When soil is compacted, the void ratio is decreased, which increases the soil's strength. Compaction does many things. Soil that is compacted will be stronger and not as likely to settle over time. In the case of foundations, settling soil can result in cracks. Therefore, compacted soil that will not be prone to settling is better than soil that is not compressed. Soil used on embankments will be more stable if it is compacted.

To most construction workers, compacting soil is simply a matter of pressing dirt together with either a tamper or a roller. In the most simple of terms, this is true. However, the compaction of soil increases the soil density by rearranging soil particles. Engineers may talk about the fracture of grains of soil and the bonds between them. You might hear the experts talk about bending the soil. Terms like "cohesive resistance" may be used. What does it all mean? It means compacting the dirt with a tamper or a roller. Sometimes water is added to the soil to improve the compaction rate. As far as most developers are concerned, knowing that compacted soil is stronger and more stable than loose soil is enough.

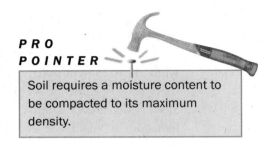

PRO POINTER
Soil requires a moisture content to be compacted to its maximum density.

Soil requires a moisture content to be compacted to its maximum density. When fill dirt is hauled in, it is usually compacted to accept weight loads. Arriving at maximum density is desirable. The moisture content required for maximum strength varies from soil type to soil type. For example, sandy soil does well with a moisture content of about 8 percent. Clay, on the other hand, compacts best with a moisture content of about 20 percent. Project engineers will evaluate fill areas and call for certain compaction specifications as needed to meet the requirements of your development.

As most builders know, soil compacts best when the soil depth is kept shallow. For example, a plumber would not backfill a sewer ditch with 3 feet of dirt and then run a tamper over it. The ditch would be filled with layers of dirt, and each layer would be tamped before the next layer of dirt was introduced into the ditch. Compaction in stages is the best way to arrive at maximum strength.

Soil that is being added to an area and compacted should be added a little at a time. Most soils should be added in layers that are not more than 8 inches thick. Each layer is compacted before the next layer is added. In some cases, water is applied to the layers of fill to increase the compaction rate. The layers are often called lifts. Some types of fill, such as gravel and sand, might be added in lifts that run up to a foot in depth. Compacting soil in layers takes time. It might seem tempting to fill an area with dirt as fast as possible and move on with other parts of the development. This would be a costly mistake. Soil that is not compacted properly is likely to cause a number of problems for a developer. Don't cut corners on soil compaction. Factor in the time and cost required to do the job properly and stick to a proven plan for suitable compaction.

Compaction Equipment

Compaction equipment options exist for developers. What is the best type of compaction equipment to use on your project? You site contractor will most likely decide on the type of equipment to be used. However, your engineers may specify the type of equipment that must be used to create a satisfactory compaction. As a developer, you should have a general idea of what type of equipment is used for various jobs. Rollers are the most common type of equipment used for large-scale soil compaction. But there are different types of rollers. Should your project be prepared with a smooth roller, a sheepsfoot roller, or a vibratory roller? You may decide to leave decisions pertaining to equipment up to your engineers and contractors, but let's take just a few moments to look at the roles of different types of rollers.

> **PRO POINTER**
>
> Soil that is being added to an area and compacted should be added a little at a time. Most soils should be added in layers that are not more than 8 inches thick. Each layer is compacted before the next layer is added. In some cases water is applied to the layers of fill to increase the compaction rate. The layers are often called lifts.

Sheepsfoot Rollers

Sheepsfoot rollers use drum wheels that have a large number of bumps or protrusions on them. The protrusions direct a lot of pressure into small areas for tight compaction. Water can be used to fill the roller drum for additional weight. Clay is compacted very well with a sheepsfoot roller. The bumps on the roller wheel can usually deliver up to 1000 pounds of pressure per square inch (psi).

Smooth Rollers

Smooth drum rollers are used most often for finish work. Since the roller drum is smooth, it maintains full contact with the soil at all times. Any type of soil except rocky soil can be compacted with a smooth roller. Since the roller drum maintains full contact, it does not create a tremendous amount of pressure per square inch. Compare a sheepsfoot roller at 1000 psi compaction to a smooth roller at about 55 psi.

Rubber Rollers

Rollers with rubber tires might not seem like much of a compaction tool, but they do work well. Pneumatic tires are spaced close together with a rubber roller. Weight is added to the equipment to obtain compaction pressures up to about 150 psi. You can get up to 80 percent coverage with a rubber-tired roller. However, this type of roller should not be used for initial compaction of some clay soils.

Vibrating Rollers

Vibrating rollers are often used in roadwork. Like other rollers, vibrating rollers compress soil with the weight of the roller, but as an added bonus the pounding of the vibrating roller packs soil even more tightly. Both granular soil and rock fill can be compacted with a vibrating roller, but the roller must be set up properly for the type of material being compacted. For example, rock fill would call for the roller to be set for a heavy weight with a low-frequency vibration. Sand, on the other hand, would call for the roller to produce light- to medium-weight and high-frequency vibrations.

Power Tampers

Power tampers are used where compaction areas are small and difficult to gain access to. The tamper is usually gas-powered and operated by one person, who walks behind the equipment. When trenches are backfilled, power tampers are most likely to be used. Since the tampers are small and fairly easy to handle, they can be put in a small trench and operated by a single person. Any type of inorganic soil can be compacted with a power tamper.

Stability

The stability of soil is directly related to its shear strength. What is shear strength? It is a rating of soil that is determined by calculating the resistance to sliding between soil particles. Physical characteristics of soil determine the shear-strength potential. For example, the confining pressure, surface roughness of particles, and soil density all affect the shear strength of a soil. Voids between soil particles are known as pore spaces. Water can flow through pore spaces and affect shear strength.

The stability of slopes must be considered carefully. Embankments fail for various reasons. The strength of any fill material on a slope can affect slope stability.

Existing soil strength must also be considered in terms of slope failure. Drainage on a slope is another factor to consider when evaluating the possible failure of an embankment. Engineers can devise methods for stabilizing most types of soil. Building an embankment is not as simple as just piling up dirt. Erosion and general failure must be assessed.

PRO POINTER

What is shear strength? It is a rating of soil that is determined by calculating the resistance to sliding between soil particles. Physical characteristics of soil determine the shear-strength potential.

Importance of Soil Assessment

The consistency of land is often taken for granted by builders and developers. People walk over a piece of land and don't often give much thought to the soil under their feet. Once development is started, the soil issues can become much more intense. A lot of money can be lost if the reading of the soil on a project is not done correctly. Developers cannot afford to skim over soil issues. Someone has to dig deeply into all aspects of soil characteristics. Search out qualified experts and let them do their jobs. Don't attempt to become your own expert in soil analysis, but learn enough about soils so that you can understand most of what your experts present to you.

CHAPTER TWELVE

Calculating Land Loss from Road Costs and Green Space

Developers soon find that roads can account for a lot of lost land. And roads can be extremely expensive to build, even in the most rural settings. If you need to build paved streets to state standards, the cost can be staggering. As you are probably starting to see, there are a lot of elements that make developing land a tricky business. There are so many ways to make costly mistakes that the odds seem to be against you every step of the way. Of course there are risks and dangers. It's easy to lose a bundle of money when you embark on a development deal. Having deep pockets helps, but profit is profit. If you are looking to make money, and most developers are, net profit is your goal. Having a lot of money to fix your mistakes can keep you out of the bankruptcy court, but it may not help you to turn a profit. To make money, you have to be smart, in tune with market conditions, and be willing to pay plenty of attention to detail.

Large developments require extensive roadwork. Major money will be on the line. Most people would expect to spend a lot of money to build paved streets with curbs and storm sewers throughout an extensive development. But would you expect to spend tons of money to put a gravel road into a country development? Don't think that roads are going to be cheap, regardless of where or how they are built. Installing a modest gravel road can cost a small fortune.

Let's assume that you are aware that roadwork will be expensive. Maybe you focus most of your energy on getting estimates for the construction and engineering costs. Are you forgetting something? You might be; many inexperienced developers do. Even if you calculate your road needs accurately, you could still come out of your development with a lot less money than you had hoped for. If the roads come in on budget, how could they cost you more money than you expected? If you did not take into consideration the amount of raw land that has to be dedicated to roads, you could lose substantial planned income due to having less land to develop into building sites.

> **PRO POINTER**
> Simple roads require a certain amount of land. Complex street systems eat up a lot of land. Far too many developers fail to plan for the land lost to road construction.

It's easy to see a 20-acre parcel of land as 20 acres of building resources. But how much of the 20 acres will be left after roads are installed? If you are planning to sell quarter-acre lots, you might think that you will have 80 sites to sell. But how many lots will you really have? Many factors consume land that you might think is available for building lots. Some examples of these needs include the following:

- Some raw land is lost to drainage needs.
- Land can be given up for recreational areas.
- Common space in a subdivision reduces land area for building.
- Roads consume a portion of your development parcel.

All these land losses must be factored in when you are estimating you total sale units. Engineers, land planners, and surveyors can help you arrive at a viable number of building lots.

Simple Access Roads

Simple access roads are usually not difficult to design. In many cases they are straight or nearly straight cuts through a property. As simple as they sound, the roads can still be extremely expensive to build. Construction cost is related to several factors:

> **PRO POINTER**
> The amount of land that you see on a plot plan will not yield as many building lots as you may hope for. In other words, factor in your land loss before you speculate on your potential number of sale units.

- Existing ground conditions are a prime consideration when it comes to cost.
- Some land is solid enough to support a low-density road without much site work.
- Other pieces of land require extensive preparation to accept a road that will last.
- Drainage is another factor in road cost.
- Distance is, of course, an element of expense for a road.
- Topography is another factor that can have a lot of influence on the cost of a road.

Some developments require extensive engineering reports. Other sites don't call for much more than a site visit by a good road contractor. Many builders are able to design their own access roads for small developments. It's wise to get engineering reports for all roads, but a lot of builders and developers skip the engineering phase when the roads being installed are simple and designed for low traffic flow.

Straight and to the Point

Keeping an access road straight and to the point is usually the least expensive way to provide ingress and egress to building lots. I've developed tracts of land that contained hundreds of acres and required only one fairly straight road. This type of road construction is about as economical as it gets. A straight run also limits land loss. By installing a road through the middle of a property, you can offer each building lot frontage on the road. All that is required is a serviceable road that has provisions for turning around at the end. If the road runs through a property and ties into other roads at each end, you don't even need the cul-de-sac for turning around. It doesn't get any easier than this.

> **PRO POINTER**
> Keeping an access road straight and to the point is usually the least expensive way to provide ingress and egress to building lots.

Branches

Some land requires road access that consists of branches built off the primary access road. Building the branches adds to the cost of construction, but the additional expense may be worthwhile. Cutting a straight road through the middle of a property might cause you to lose development potential. You may be able to get more building lots out of a parcel if you install branch access roads.

Branch roads give a developer the opportunity to have more lots fronting on a road with less land. Yet the branch roads require land. Is the land lost to road construction a waste? It depends on the project. The shape of a land parcel comes into play with road design. Zoning requirements and desired lot sizes are also factors to consider when thinking of branch roads. Many subdivisions must incorporate the use of branch roads to reach the highest and best use. Your land planner and engineers will be your best sources of suggestions for road layouts.

Natural Road Sites

Natural road sites exist on some land. If you have walked much acreage, you've seen natural paths that lend themselves to road construction. The paths may be the result of old logging roads that were used many years ago. Sometimes the road sites are simply a result of topography. Almost anyone can spot some of the natural road sites. What are you looking for? Solid, high ground is a good start when looking for a road location. If there are gaps between trees or only small brush growing in a potential path, it's a plus. Negative factors for a roadway can include:

- Hills
- Streams
- Low spots
- Soft ground

When natural road sites exist, you should try to use them. Building a road, even a simple one, is not a cheap proposition.

Complex Roads

Complex road systems must be engineered. If you are going to do a development that will have miles of paved roads, you would be extremely foolish to start construction without engineering reports. Some developers feel that having road contractors come to a site and make suggestions is enough. I strongly suggest that you pay for engineering reports for any major roadwork. The difference in cost between building a gravel road through some farmland and installing a paved road system in a large subdivision is immense. Mistakes made with paved roads are considerably more expensive than the same mistakes made with a gravel road. Don't attempt any major roadwork without plans from experts.

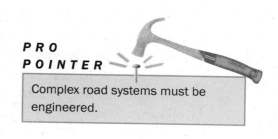

PRO POINTER

Complex road systems must be engineered.

Typical developers don't delve deeply into the road-planning process for large projects. Yet a good developer should be aware of the amount of land that must be used for:

- Roads
- Curbs
- Parking areas
- Turnarounds

These needs are often taken for granted, but they shouldn't be. The amount of land required for road construction, parking, and related features can be substantial. Losing land is like losing money. It is the land that you will be selling; you won't be selling roads and parking lots.

Good designers can come up with road plans that are cost-effective and efficient. It is not wise to cut corners on design issues that may repel land buyers when your development is complete. To expand on this, let's talk briefly about your options for turnarounds on dead-end streets and how decisions pertaining to the turnarounds could affect your development.

When you think of a turnaround area, what do you envision? Most people from urban areas think of cul-de-sacs. The use of cul-de-sacs is widespread and accepted as something of an industry standard. But, there are other, less expensive types of turnarounds that could be used. How much money should you try to save? Reducing development costs is a great way to increase profits. But if the cuts you make reduce sales or result in lower sale prices, they will prove to be frustrating.

Let's assume that you want to use cul-de-sacs for your turnaround areas. It is your desire to use island circles, which is a type of cul-de-sac that is circular in shape and that has an island inside the travel area so that landscaping can be planted. This is a classy type of turnaround, but it is expensive. You weigh the cost and decide to do without the island feature. The decision will save you some money, and you will still have a circular cul-de-sac for your turnaround. A decision like this may be quite sensible. But, what are your other options?

PRO POINTER
You can go too far in trying to save money on road construction. Trying to avoid engineering reports can be a bad mistake.

Small developments that have low traffic flow can get by with T-shaped turnarounds. A T-shaped turn is not as easy to turn around in as a cul-de-sac. Due to the design of a T-turn a driver must pull into it, pull forward in one direction, back up to the other end, and then reenter the street. This takes time and can be difficult for some drivers to deal with. The amount of land and pavement needed for a T-turn is less than that required for a cul-de-sac, so money can be saved. But are you buying trouble for your development? Saving money on the turnarounds could cost you in lost sales or unhappy residents.

A Y-shaped turnaround is similar to a T-turn except that the angles are easier to negotiate. More land and pavement are needed for a Y-turn, but it requires less land and pavement than a cul-de-sac. So what should you do? Most developers would be willing to spend more for cul-de-sacs to ensure acceptance of their developments by land buyers. However, if you were dealing with a small development of, say, six houses on a gravel road, a Y-turn would probably be fine. Generally speaking, cul-de-sacs are best.

Failing to make the roads user-friendly will prove to be a problem. Being unwilling to spend a little extra money to make roads more attractive can hurt you. Major design issues will be handled by your experts. Your role in road design may not be large, but at the very least keep in mind that all the areas used for roads and related needs will deplete the amount of raw land that you will have left to sell.

Green Space

Green space is going to eat into your land holdings. Don't think of this as a loss. Think of it as an advantage. Creating a green development will give you reason to ask for a higher sales price on your lots. People who want to live in a green world are generally willing to pay for the privilege.

How much land will be set aside for green purposes? This is largely up to you and your land planners. There are many ways to maintain a sustainable development.

Some of what you reserve for green purposes will be environmental. Other elements will be enhancements for residents:

- Natural walking paths
- Bird-watching areas
- Earth-friendly playgrounds
- Fishing ponds
- Natural exercise trail stations
- Botanical gardens
- Wildlife feed patches
- Living-history displays that identify trees and similar items

In addition to the lifestyle green space, you will likely lose land to greener landscaping and drainage techniques. This could come in the form of sloping land, retainage ponds, pervious parking, and associated elements.

All of your proposed land loss should be calculated before you jump into deep water. It is a common mistake among rookie developers to look at total acreage and assume that each acre will result in building lots. This simply is not the case. There will be land loss during the development phase. To be profitable, this loss must be factored into your pricing structure.

CHAPTER THIRTEEN

Water Requirements

The water requirements for a development must be calculated in order to arrive at a cost projection for a project. Figuring the water requirements for a sizable development is a job for experts. Even small developments can have diverse water needs that should be designed by experts. There is more to water needs than just the water used by homes in your development. Irrigation is a need that some investors might overlook. Having water available for fire hydrants may be a necessity. If there are community buildings in your development, water will be needed for them. Some developments contain homes in which fire sprinklers are standard equipment. Fire-protection systems can require greater water pressure and quantity. In short, the need for water in a development involves much more than some developers realize.

Some parcels of land don't have access to city water mains. In some cases entire communities run on well water. Community well systems can be expensive and unpredictable. Most developers look for land where water is readily available. Buying land where well water will be needed can be risky. But well water is used in many places without problems. I've done a lot of small developments where the only water came from wells. Whether you are counting on municipal water or well water, you should not take anything for granted.

I've made some expensive mistakes in dealing with city water mains. One development I handled had a city water service running in front of it. I knew the water main was there and assumed that all I'd have to do was pay the city tap fee and connect to the water main. I was wrong. The water main was there, but it was too small to serve my little development. I was required to upgrade the water main and repair the street cuts out of my development profits. Ouch!

I've done jobs where I thought that the city would extend water service to my property at the city's expense. Cities often do this, but I ran into a situation where the city required me to do my own street cuts and repairs. Repairing street cuts to meet demanding state and local requirements can be very expensive. It only took one experience to teach me to ask for complete details before assuming anything about city policies.

> **PRO POINTER**
>
> The cost of installing water systems can be staggering. Developers who work with large projects should rely on experts to plan water systems. Builders and developers who are working on very small developments may not need any engineering help, but they still have to cover their bases.

The cost of installing water systems can be staggering. Developers who work with large developments should rely on experts to plan water systems. Builders and developers who are working on very small developments may not need any engineering help, but they still have to cover their bases. Wells for individual building lots are one matter, but any connection to a public utility is completely different. Developers need to identify the water requirements of their projects early in the planning stage.

Types of Water Demand

There are two types of water demand—consumptive and nonconsumptive use. Water that is needed for municipal, agricultural, or industry use is considered consumptive. Nonconsumptive use involves recreation, transportation, and hydropower. Developers are most often concerned about the water requirements for the following uses:

- Institutional needs
- Residential needs
- Commercial needs
- Industrial needs
- Recreational needs
- Firefighting needs

- Irrigation needs
- Decorative needs

There are charts and tables available to help designers determine the water needs for a development. Most plumbing codes provide such tables. The amounts of water required by local codes can be somewhat shocking. Most people are not aware of how much water may be needed for a single household. For example, you may find that the minimum required daily water rate for a residential development equates to 100 gallons of water per day per person. If you were planning for a shopping center, you might have to figure on up to 300 gallons of water per 1000 square feet of floor space. Designers normally handle the calculation of water requirements for developers, so you should not have to become personally involved with the process.

When water systems are sized, they must allow for the delivery of enough water to cover peak demands, fire flows, and increased future demands. Most of the time a system will perform at a level far below its capacity. But it must be large enough to meet all demands when necessary. In addition to peak demand, water pressure has to be considered. Again, normal use and fire flow must be factored into the sizing. Since water mains are usually installed below ground and follow the topography of land, more pressure is needed to push water over hilly terrain. In order for a water main to have adequate pressure, it may be necessary for individual buildings to be equipped with pressure-reducing valves. For example, a residence may be limited by the plumbing code to a maximum working pressure of no more than 80 pounds per square inch (psi). If a water main is delivering pressure at a rate of 125 psi, a pressure-reducing valve will have to be installed in the home to lower the pressure to an acceptable level.

Local authorities normally set minimum requirements for water-pressure needs. State agencies often handle this part of the design function. Fire flow requirements are usually established by individual municipalities. The requirements for minimum pressure in water mains must be reviewed when sizing a system for a new development. Again, this work should be done by your experts.

> **PRO POINTER**
>
> The routing of water mains can require considerable planning. Depending on where you will be developing land, you may find that specific routing requirements for your water system are mandated. Jurisdictions often set guidelines for the placement of water lines.

Routing Water Mains

The routing of water mains can require considerable planning. Depending on where you will be developing land, you may find that specific routing requirements for your water system are mandated. Jurisdictions often set guidelines for the placement of water lines. For example, you can expect that a rule is in place to keep all water pipes at least 10 feet away from all sewer pipes. This rule protects the pipes in case one or the other (or both) develops leaks. When water mains are allowed closer than 10 feet to a sewer, the water main is likely to be required to be installed on an elevated shelf that is at least 18 inches above the sewer. Additional safeguards may be required to prevent contamination of the water system.

Actual placement of water mains can be restricted to certain locations. For example, a city may require that all water mains be placed a fixed distance from the centerline of streets or curbs. You may be required to install all water mains on only one side of a street to maintain conformity. I've heard of situations where cities were very strict about where water mains could be installed, so this is an issue that must be confirmed. The experts who are designing your project should take care of this for you.

Depending upon your designers' plans, you may have to create easements for the installation of your water system. If easements are required, you need to create them before you start selling lots. Developers try to keep easements on outparcels and common areas. Whenever possible, avoid running water mains through lots you plan to sell. If you do have to encroach on individual lots, keep the water mains as close to the property lines as possible.

Residential developments don't pose as many problems in the placement of water mains as commercial sites do. However, high-density residential developments can be as difficult to work with as commercial sites. What makes commercial projects and high-density projects more troublesome? Space is generally very limited with commercial projects. The same can be true for townhouse and condo projects. When there is limited land to work with, the placement of utilities can be very difficult.

Due to the size of water mains, they can require substantial room. Small residential developments might be served by a water main with a diameter of as little as 4 inches. However, a 6-inch diameter is generally considered about the smallest size suitable for residential

> **PRO POINTER**
> If you will need community space to accommodate development needs, consider establishing easements and right-of-ways prior to marketing your building lots.

projects. Large projects may require water mains with diameters of 24 inches or larger. Routing sewers and water mains can be a bit like running a maze. Designers can really have their work cut out for them if there is limited land to work with.

Developers must consider the quality of the land where they plan to install utilities. Having enough land is one thing; having enough land that is suitable for an installation can be quite another. For example, I could buy 60 acres of land here in Maine and be hard-pressed to find a suitable path for water and sewer lines. How could this be with so much land? Bedrock could make installing pipes below grade extremely difficult and expensive.

> **PRO POINTER**
> Beware of underground obstructions, such as bedrock, that may cause problems with the installation of utilities.

Sewers and water mains are normally buried. Minimum ground cover for a small sewer is usually 12 inches. Water mains should be installed below the local frost line, which could easily range from 20 inches to 48 inches or more. Can you imagine what it would cost to create a path through bedrock for a water main that had to be buried 48 inches deep? I have run into bedrock when running water services to individual homes. In such cases, I got minimal bury depth and installed in-pipe heat tape to protect the pipe from freezing during cold temperatures. This is feasible for a single water service for a house, but it certainly wouldn't make sense for a development. If a water service and a sewer are installed in a common trench, the water pipe must be above the sewer and on an independent shelf within the trench to reduce the risk of contamination if the pipes were to leak.

Part of the design process for routing water mains involves valve, fire-hydrant, and connection locations. Determining placement for fire hydrants is a job that must be done with careful attention to fire codes and local regulations.

Fire Hydrants

Fire hydrants are not installed in all developments. Some areas simply don't have the resources for fire hydrants. But many developments are required to be equipped with hydrants for fire fighting. In these cases, designers must work

> **PRO POINTER**
> Many developments are required to be equipped with hydrants for fire fighting. In these cases, designers must work within the confines of established rules.

within the confines of established rules. There are some basic guidelines that might be followed, but it is more likely that there will be specific regulations to observe. Since rules vary, let's talk about the basics.

It's obvious that fire hydrants have to be readily accessible and placed in conspicuous locations that are close enough to buildings to make the hydrants worthwhile. A few developers resist scattering fire hydrants all over a development. The view of these developers is that the hydrants detract from the aesthetic quality of a development. If you share these feelings, get over them. You have to make fire-fighting equipment readily available. Your designers may be able to come up with some creative ways to reduce the potential negative impact on the appearance of your development.

It is common for fire hydrants to be installed within a few feet of the edge of a street or curb. Connections on the hydrants should face the street. Precise locations should take into consideration the ease or difficulty with which firefighters will be able to connect to the hydrants. You may find that hydrants must be installed a certain distance from buildings. There are two reasons for this. Hydrants must be close enough to buildings to make them effective, but the hydrants must be far enough from buildings that might be burning to make it safe for firefighters to connect to the hydrants.

The rules for hydrant placing can vary with the type of development that you are involved in. Typical residential developments might require different hydrant placement than what a townhouse project would. Commercial developments are certainly likely to require different placement for hydrants. Some commercial buildings have manifold hookups on exterior walls so that fire hoses can be connected directly to the building. When this is the case, the manifold is usually kept within 100 feet of a fire hydrant. The spacing between fire hydrants varies. It is common for hydrant spacing to run between 300 to 1000 feet. Your designers, along with the help of local regulations, will spot the locations for fire hydrants. Good designers will take into consideration the need for visual appeal and will keep obtrusive locations to a minimum.

Tapping In

Tapping into existing water mains can be very tricky. Some existing water mains cannot be shut down even for short periods of time. When this is the case, the tap-in requires special equipment and skill. In simple terms, two halves of pipe are placed over the

PRO POINTER

Tapping into existing water mains can be very tricky. Some existing water mains cannot be shut down even for short periods of time. When this is the case, the tap-in requires special equipment and skill.

existing water main and strapped into place. The strapping process involves bolting together the flanges on the pipe halves. A cutting tool that fits over a valve cuts into the water main through the open water valve. Once access is gained to the main, the cutter is removed and the valve is closed. When all goes as planned, the remainder of the work is dry and typical.

When isolation valves are available, an existing water main can be shut down long enough for a dry tap to be made. While the hookup is being connected, water to buildings downstream is turned off. This is the simplest way of tapping in, but it is not always practical. Wet taps are often required, and they tend to be more expensive than dry taps. This is an issue that you should investigate while putting together your numbers for the cost of a development.

There may be an existing water main available for your project. Sometimes each building can have an individual tap into an existing water main. Other circumstances require one large tap-in that is developed into a complete distribution system within a development. Either way, there will be tap fees, and they can cost thousands of dollars.

Wells

Water wells are sometimes used to provide water for developments. This might involve a community water supply that is distributed through a water-main system or individual wells for each building. Individual wells are fairly simple. Community water services are not really complicated, but they can require several pumping stations and holding tanks in addition to normal distribution piping. Cost is certainly a factor in determining which type of well system to use when municipal water mains are not available for tapping.

Some homebuyers are reluctant to buy houses if their domestic water comes from a well. This is something that you must consider when doing your marketing research. There are many areas where wells are common and meet with no resistance. However, if you are developing in an area where most homes are served by municipal water mains, you may find it difficult to sell your building lots. Research this aspect carefully before you commit to a development.

When a community well system is used, it is likely to be made up of several wells and pumping stations. Large holding tanks are needed for community well systems. Ongoing maintenance of the system may also be a cost factor. I have seen developments of this type, but I've never undertaken one myself, and I wouldn't. I think that the process is too complicated to deal with, but that's only my personal preference. However, I have done many projects where individual wells were used.

Individual wells are usually either drilled or dug. Dug wells are common in the south, and drilled wells are the norm in the north. Drilled wells are more dependable, but they are also more expensive. Most developers who install wells use whichever type of well is most common in the local area. I've installed both types of wells. A few of the dug wells, also known as shallow wells, came back to haunt me. They sometimes went dry during hot, dry summer months. A few of the plumbing systems served by the wells sucked sand into the water lines. I have never experienced any problems with drilled wells. So my vote goes to drilled wells, but both are suitable subject to local conditions.

If you are dealing with individual wells, you avoid the cost of expensive distribution systems. However, the cost of the wells and related well equipment, such as pumps and pressure tanks, adds up quickly. Usually, if a public water main is available, a developer must use it. It may be cheaper to use wells, but most jurisdictions will not allow the use of private water or sewer systems when public systems are available. Clearly, this issue is something that you must address fully before cementing your costs for construction. I strongly suggest that you have your experts look into the matter for you.

CHAPTER FOURTEEN

Flood Zones, Wetlands, and Other Deal-Stoppers

Flood zones, wetlands, and other deal-stoppers can be a developer's worst nightmare. There are simply some elements of land that can't be worked around. Developers who buy land without adequate research can wind up in deep financial trouble. I guess I've been lucky. While I've come very close to being devastated by land problems, I've never taken a direct hit. Even so, I have lost money that I hadn't planned on spending due to land elements. Developers who deal with environmental issues have to be able to take a lot of heat and maintain their cool.

Most developers shy away from anything that might be close to an environmental issue, but there are some who have no fear. Then there are those who are willing to walk the line and hope that they don't get too close to the edge. Over the years I have done my share of line walking, and I've known some developers who have shown total disregard for environmental laws.

Personally, I have a great respect for environmental issues. There isn't enough money to make me destroy natural resources that should be protected. When I say that I've worked the edge, I simply mean that I've built in areas where I've taken heat and been proved to be in the right. The last house I built for myself is an example of such a situation, and it's a good example of how even careful builders and developers can wind up in hot water, so let me tell you a little about it.

I bought 25 acres of land to build my personal home. The 7 acres on which I chose to build fronted on a small river. The river was in view of the home location, but I couldn't see any flood risk or wetlands issues. However, to be safe, I asked the

Department of Environmental Protection (DEP) to inspect the site. A representative from the DEP came out and assured me that there was no threat to the river or the wetlands from my proposed construction. My next step was to have the local building inspector come out for a site visit.

The local building inspector assured me that the land was fine to build on and that getting a building permit would not be a problem, subject to a soils test. I had a soils test done and it was fine. Everything seemed okay. Once I was convinced that the deal was safe, I removed all contingencies from my purchase contract and bought the land. A few weeks later, when I applied for my building permit, the permit application was denied. The reason I was given was that the lot was not a legal building lot. I was told that it was an illegal subdivision of a larger parcel. A few weeks before I had been told that the lot was fine, and all of a sudden I was in a world of trouble.

Without detailing all the steps required, I went to the zoning board. The next step was the local board of appeals. My attorney and I had done exhaustive research to prove that the town was wrong in its opinion of the land status. Long story short: I won my appeal and got my building permit. I'd done everything reasonable to make sure that the parcel was an approved building lot and still ran into roadblocks. Due to my experience, knowledge, and persistence I won. If I had been an average homebuyer I might have given up.

After the fiasco of the building permit I thought that my troubles were over, and they were for a while. When I financed the home, I got a construction loan that would convert to a permanent mortgage. The house was finished and the permanent mortgage went into effect. After several months I decided to refinance the loan for a better interest rate. When I did, another problem came up. The bank told me that the house was built in a flood zone. I was shocked and didn't believe it. A survey crew was sent out by the bank.

The bank's survey crew told me that the house was "probably" in the flood zone. I looked into flood insurance and found that, since the town where my home was located didn't participate in the flood program, I could not get insurance. I was steamed. Then I found out that I couldn't sue the town, since it was not insured and a legal tort protected such towns. Things were going from bad to worse.

> **PRO POINTER**
>
> I've seen tiny frog ponds kill development deals. In fact, I can remember a piece of land where the mere presence of ferns and cattails scared off a major developer.

My new loan was in jeopardy. Both the new bank and the old bank were on my case. I hired a surveyor who just happened to be on the town council to do a full survey not only of my land but of the entire river area. It was very expensive, but it proved that my house was not in danger of flooding. This resolved the issue for good. My expense was considerable, and the mental anguish was extreme, but I won. All this fighting should have been unnecessary. The land I bought was not a problem piece, but it turned into one.

As you can tell from my personal story, even experienced builders and developers can do all that is reasonable and still wind up in a mess. You can just imagine what could happen to a developer who was careless. Environmental issues are a big factor in land development.

Working with Wetlands

Working with wetlands is a high-risk venture. Any land containing even a small section of wetlands is a potential time bomb for a developer. I've seen tiny frog ponds kill development deals. In fact, I can remember a piece of land where the mere presence of ferns and cattails scared off a major developer. It doesn't take much to put a piece of property under the scrutiny of environmental regulators.

> **PRO POINTER**
>
> One of the laws that impacts land developers most is the Emergency Wetlands Resources Act of 1986. This law ensures the conservation of wetland resources. Any wetland area can fall under multiple laws.

Several laws pertaining to environmental issues are on the books. There is a law that prohibits unauthorized obstruction or alteration of any navigable water. The law keeps developers from filling in such waters as well as preventing the excavation of material from the waters. This is a serious situation, and most developers respect it. Remember when I mentioned earlier that I knew of developers who had no respect for laws? This particular law was violated by a developer I used to know. The developer filled in wetland areas for development. He did the fill fast and knew that he would be caught.

I had lunch with him after the fact and we discussed his actions. At the time he was in deep legal trouble. His attitude was that it was his land and that he would do what he wanted with it. He told me that he had filled the wetland quickly so that the authorities would have to make him remove the fill, rather than stop him from placing the fill. The last I heard, the developer was juggling court dates and fighting his fight. Personally, I don't agree with what he did, but he did it.

Another law on the books has to do with the risk of discharging pollutants into navigable waters. There is a law that deals with the transportation of dredged materials headed for disposal in an ocean. One of the laws that impacts land developers most is the Emergency Wetlands Resources Act of 1986. This law ensures the conservation of wetland resources. Any wetland area can fall under multiple laws.

Any developer wishing to fill in a wetland area must apply for a permit. The permit application will be reviewed by both the U.S. Army Corps of Engineers (Corps) and the Environmental Protection Agency (EPA). Getting approval for such a fill request is unlikely. To obtain approval, you must demonstrate that you have no practicable alternatives to filling the wetland area. Further, you must prove that your fill will not cause significant damage to the aquatic ecosystem. The EPA has veto power over the Corps in such matters.

What is a wetland? You could probably get many answers to this question, but the definition given by the governing bodies is the definition that matters most. According to the environmental authorities a wetland area is an area that is inundated or saturated by surface or groundwater at a frequency and duration sufficient to support, and that under normal circumstances does support, a prevalence of vegetation typically adapted for life in saturated soil conditions. Wetlands generally include swamps, marshes, bogs, and similar areas. What you have just read is the official definition of a wetland, but don't assume that the definition given is all that there is to the matter. The interpretation of the definition can be broad, so you must be cautious.

Wetlands may be regulated by state, federal, and local agencies. Passing muster with one agency doesn't exempt you from the others. If you have any reason to believe that your project might fall into a wetland classification, you

> **PRO POINTER**
> Any developer wishing to fill in a wetland area must apply for a permit.

> **PRO POINTER**
> According to the environmental authorities a wetland area is an area that is inundated or saturated by surface or groundwater at a frequency and duration sufficient to support, and that under normal circumstances does support, a prevalence of vegetation typically adapted for life in saturated soil conditions.

need to involve experts to remove any doubts or risks that might jump up in your face. If you violate a wetland regulation, you may be subject to either civil or criminal action. Trying to beat or cheat the system simply isn't worth the price you may have to pay.

> **PRO POINTER**
> If you violate a wetland regulation, you may be subject to either civil or criminal action.

There are many rules that apply to the disturbance of wetlands. If you have plans for clearing land, dredging areas, or filling sections of your development, you will certainly trigger wetland regulations if the land falls under the wetland protection. Other activities can also put you in harm's way, so it's wise not to consider a project in any area where you might be nailed for a wetland infraction.

Flood Areas

Flood areas are bad for developers. The benchmark for flood areas is usually the 100-year flood boundary. This is an area of land that has been flooded within the last 100 years. Most communities participate in a flood program that allows homeowners to acquire flood insurance at reasonable rates. Even at the reasonable rates, flood insurance is an expense that is not required for properties that are not considered to be at risk of flooding. If you create a development where flood insurance is required, selling the lots could be difficult. On the outside chance that you get caught up in a deal where flood insurance is needed but not available, you are in deep trouble.

> **PRO POINTER**
> If you create a development where flood insurance is required, selling the lots could be difficult.

Local authorities normally have flood maps available for inspection. However, the maps may be old and difficult to read. This is the problem that I ran into with my personal home. The flood maps were old and were not drawn in great detail. It took a detailed survey, which I had to pay for, to change

> **PRO POINTER**
> Have your engineers establish local flood areas and make sure that your building sites are not in them.

the minds of the local authorities. Have your engineers establish local flood areas and make sure that your building sites are not in them.

Hazardous Waste

Hazardous waste is a component of modern land developing that old-time developers didn't have to worry so much about. Times have changed, and hazardous waste is a serious consideration in modern development practices. The EPA offers a list of materials that are considered to be hazardous. Materials not listed may also be considered to be hazardous if they exhibit any of the following characteristics:

- Toxicity
- Ignitability
- Corrosivity
- Reactivity

If you become involved with hazardous wastes, you may have a lot of hoops to jump through. The most innocent piece of property can harbor hidden waste and high cleanup costs. If you violate regulations pertaining to hazardous wastes, you may have to foot the bill for all cleanup costs, which will not be cheap. In addition to the cleanup costs you could be hit with fines of up to $25,000 for each violation.

There is no way that I know of to be absolutely sure that a parcel of land might not be hiding hazardous waste. Research of past use of the land is about the best defense that you have. It is possible to do expensive scans of the property, but if general research doesn't raise any red flags there is probably no need for high-tech scans.

Other Environmental Concerns

Other environmental concerns for developers could range from destroying natural habitat for an endangered species to erosion. The list of potential risks can be a long one. This is why you need to bring in an environmental expert to clear your project before you go too far in starting the developing process. Spe-

> **PRO POINTER**
> Other environmental concerns for developers could range from destroying natural habitat for an endangered species to erosion.

cialists can be expensive, but they are a real bargain if you compare the risks and costs of what could happen without them.

So far we have talked about hard-line environmental issues. The subjects covered up to this point are dealt with under some form of legal protection. But there are other issues that you could face as a developer that are not so clearly defined. Sometimes a developer's worst enemy is the public, and you might find yourself in some situation where public disapproval is the biggest drawback to a development.

Angry Citizens

Angry citizens can be extremely difficult to deal with. Your development plans can meet all requirements, be approved, and still fall into a public-relations trap that is hard to emerge from. I've had this happen on a small scale, and I've known other developers who have dealt with the problems on a much larger scale. People don't always embrace new developments. While there may be no legal reason or way for the public to stop your development, people can certainly make the profitability of your project suffer. Let me give you a few quick examples from my past that will highlight this facet of developing.

> **PRO POINTER**
> If the people near your land don't support your development plans, you could run into expensive vandalism.

I bought some leftover lots in a subdivision to build on. The subdivision was several years old, and residents had become accustomed to using the vacant lots for their own purposes, such as mulch disposal. When I bought the lots, there was some distress that the vacant lots were about to be built on. One neighboring resident was especially nasty about my company building in the area. The resident was rude to my workers and would not cooperate with us in any way. When I had surveyors stake the house out so that footings could be dug, the survey stakes came up missing the next day. I paid the surveyors to stake out the house again, and again the stakes were gone when the backhoe operator arrived the next morning. Finally, I paid a third time to have the house staked off and had the surveyors drive iron stakes below the ground level. The next morning I was on the site with my metal detector and found the stakes for the house corners. This problem cost me a few days and a few hundred dollars more for extra surveys, but I finally got the footings in. All in all, compared to some horror stories, my experience wasn't too bad.

I have known of developers who encountered many problems with developments due to unhappy neighbors. The types of problems ranged from people pouring sugar into the fuel tanks of heavy equipment to windows being broken in houses on a regular basis. If the people near your land don't support your development plans, you could run into expensive vandalism. While I have no first-hand knowledge of anyone forming a blockade or a physical protest on a project, I have heard of such events.

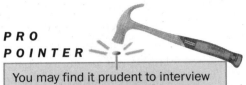

> **PRO POINTER**
>
> You may find it prudent to interview residents in the adjacent areas of your development to see if there will be any mass disapproval of your intended plans that could lead to expensive remedies.

If your project is approved and legal, you have remedies against people who stand in your way, but that might be of little help in the real world. Calling the police daily or filing lawsuits robs you of development time. You may find it prudent to interview residents in the adjacent areas of your development to see if there will be any mass disapproval.

Land development is a business that can be plagued with problems. It is probably impossible to avoid all potential problems. However, with enough knowledge, experience, and research you can dodge most of the bullets. There are plenty of traps waiting for the unsuspecting developer, so keep your guard up and cover all the bases that you can. Some problems are sure to slip through your defenses, but you can head most of them off with proper preparation.

CHAPTER FIFTEEN

Location, Location, Location

Real-estate professionals have long had a saying about the most important element in determining the value of real estate. It is said that the key to high value is location, location, and location. There is a lot of truth in the adage. Many old-time investors say that they try to own the worst properties in the best neighborhoods. Land developers can prosper by cashing in on the right locations at bargain prices. But don't be fooled into thinking that every low price equates to a bargain. Bad land at any price is not a good buy. Developers have to weigh many factors when deciding what type of land to acquire and where to buy it.

The type of development that you wish to create will determine to a large extent the type of land that you will need. Your type of development will also dictate the various types of locations that may be suitable for the project. Land that is ideal for a medical complex might not be worth considering for a shopping center or a residential subdivision. Prices for land vary greatly depending upon location, zoning, and other factors. Sometimes the best land bargain may be an expensive piece of property. Do you really care how much the land costs if you make a terrific profit? You could get in as much or more trouble by purchasing cheap land as you would by investing in prime property at a premium price. However, you must work the numbers for each deal to determine what you should do.

The first step in choosing your location is determining what your land needs are. If you plan to build a series of do-it-yourself car washes, you will want a location that is visible and where there is enough vehicular traffic to warrant the construction.

Developing land for a group of private storage buildings could require a different type of land and location. The storage facility might be able to be built on land that doesn't have access to water and sewer services and that will not perk for private waste disposal. Land without water and sewer facilities would not work for a car wash, but it could work for the storage facility.

If your interest is in creating office space, you will most likely want land in an urban setting. Rural land might support some office space, but being closer to the city would probably be smarter. Much of the decision-making process is a matter of common sense. Some of it is based on creative use of land. Money is always a consideration. And developers must know their market; this is where demographic studies come into play.

Would you be better off building a complex of high-end condos near a subdivision of single-family homes or in an area where other condos have been built? More information is needed to answer this type of question. You can make a decision based on your personal opinion, but if you want to make money consistently, you have to perform plenty of research. Getting to know your customers before you present them with something to buy is a major step towards becoming successful.

Matching Up Your Needs

Matching up your needs with available land is not difficult. Once you know what your land needs are, matching them to land that is available is simple. But what happens when you find three parcels that all fit your criteria? How will you decide which parcel is your best bet? Reviewing all available data on each parcel is the best way to refine your buying decision. Let's look at a short example of what you might face in a three-way comparison.

Assume that you want to build a development of Victorian homes. You want each home to be situated on a minimum of a quarter-acre lot. According to your research, the homes should have a minimum of three bedrooms and at least two bathrooms. Your plan calls for some larger homes that will have four bedrooms and two and a half baths. The plan that you have in mind calls for a semicircular layout. The street will be horseshoe-shaped and will serve all the homes. Your plans are drawn, and you want enough land to build at least 12 homes. Considering lot sizes, the subdivision street, and other land needs, you have determined that you need at least seven acres of land. You know that you might get by with a little less acreage and you wouldn't mind having a little more, but your plan indicates that seven acres is likely to be the best size for your development.

After doing some research, you find three parcels of land that might work for your development. All three parcels have water and sewer hookups available. The size of each parcel is within your guidelines, but Parcel One has more rolling land than the other two parcels do, and this could increase the development cost. Parcel Two has very few trees, and this could reduce the value of the building lots. On the other hand, the lack of trees would reduce the cost of clearing the building lots. So far, Parcel Three seems like the best choice.

The prices of all three parcels are similar, with Parcel Three being the most expensive by about $5,000. Parcel One is the cheapest piece. In terms of land for development, the price difference between the parcels is not really enough to influence your decision. You know that you will have to make your buying decision based on other factors.

> **PRO POINTER**
>
> Matching up your needs with available land is not difficult. Once you know what your land needs are, matching them to land that is available is simple. But what happens when you find three parcels that all fit your criteria? How will you decide which parcel is your best bet? Reviewing all available data on each parcel is the best way to refine your buying decision.

You will be building Victorian homes for families. Most homebuyers will have or be planning to have children. You can use demographic reports to profile your buyers. This prompts you to look into the proximity of schools and playgrounds. You also check to see what types of medical facilities are nearby. Then you check out the daycare facilities. Parcel Three, your first choice, doesn't fare well compared to the other two parcels in terms of kids' issues. Parcel One appears to have the most convenience to offer parents. Parcel Three would be great for retirement housing, but there is not much in the area to offer children and their parents. This causes you to rule out Parcel Three.

Now you are down to two parcels of similar price. One has rolling land, and the other has very few trees. When you look more closely at the two parcels, you notice something about Parcel Two that you had not paid attention to before. With your horseshoe street you need enough road frontage to connect to the main street in two places. As you scale out the road frontage and compare it to your needs, you see that there is not enough road frontage with Parcel Two. This new information kills this deal and points you back to Parcel One.

After more consideration you decide to buy Parcel One. The rolling land will make construction a bit more expensive, but the appearance of the homes will benefit from the different elevations. More importantly, the parcel fits all your needs for success. As it turns out, Parcel One was the least expensive piece of land you were considering and it was also the best piece of land for your needs. Sometimes things work out this way.

You don't have to buy the most expensive land on the market to get good land. But you must be willing to pay enough to ensure that you are getting land that will give you good odds for success in your development plan. Demographics can be a big help in deciding which land locations are suitable for your development. The example above showed the impact of demographics. Knowing that your homebuyers either had or would have children helped you to pinpoint your land needs better.

Demographics

Demographics are statistical studies of populations. A demographic study can be fairly simple, such as the number of males and females in a town, or quite complex. Items that might be determined with a complex study might include any of the following:

- The age of the population
- Income
- Family status
- Level of education of potential buyers

Let's say that you are planning to build a community of condominiums. How many bedrooms should the units have? Rule-of-thumb planning would indicate that three-bedroom units should attract the most interest. This may be true, but will your target market be able to afford three-bedroom units? You need to know what your buyers are

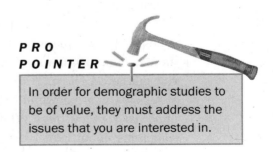

PRO POINTER

In order for demographic studies to be of value, they must address the issues that you are interested in.

likely to want and what they are probably going to be able to afford. It might be determined that your complex of condos should offer one-bedroom units at lower prices to coincide with your buying public. Maybe you should offer loft units as an option. To know what to build, you have to know what people in the area will want and what they can afford to pay. This type of information comes from demographic studies.

Demographic studies may be developed by local or state agencies. To get refined, detailed studies, you may have to hire a research firm to conduct a survey specific to the information that you are in search of. Looking around an area might be all that you have to do. By observing the cars, the age of people, and the other housing in an area, you can make assumptions. But we all know that assumptions can be dangerous. Developers who are working with large projects can afford to pay for detailed, personalized studies. Small developers may not have a budget for such a study.

PRO POINTER

Demographic studies may be developed by local or state agencies. To get refined, detailed studies, you may have to hire a research firm to conduct a survey specific to the information that you are in search of. Looking around an area might be all that you have to do. By observing the cars, the age of people, and the other housing in an area, you can make assumptions. But we all know that assumptions can be dangerous.

If you are on a tight budget and can't afford a customized study, you still have options. Check with your local town or city offices to see if any census or demographic studies are available. Contact real-estate agencies and see if they will share any demographic information with you. Real-estate appraisers may also be able to provide you with demographic data. If you are working on a project that is noncompetitive with other developers in your area, you might get some demographic details from the other developers. While some of these sources may charge you something for the information, the cost will probably be much lower than it would be if you commissioned your own study.

In order for demographic studies to be of value, they must address the issues that you are interested in. For example, if you are planning to create a golf-course community, you will be investing a lot of money in the course. The recreational opportunity should add considerable value to the development, but will people in the area be interested in living around a golf course? Will they be able to pay the higher cost for such housing? You need to know the answers to these questions before you build a golf course. A demographic study can give you the answers.

Let's say that you are thinking of developing land for professional office space. Will the area support the new offices? Researching this type of question can be easier than figuring out if a golf course is a viable amenity to a housing project. Get on the phone and call rental agencies. See if they have regular requests for professional office space. Ask if there is more demand for space than there is space available. Don't rely completely on the results of your phone calls, but keep notes on your findings. You can try other types of research to complement your phone calls. For example, you can run a few ads in the local newspaper offering free information on a new development of professional office space. If you get very few responses, you should be nervous about your plans. But if you are flooded with requests for brochures, you should feel pretty good about them.

Straight demographic reports are easy to interpret. Most small developers use many types of studies to determine how well their developments might work out. Your engineers should be able to provide you with substantial demographic data. I've hired people to sit down for hours at a time and make phone calls to interview people about what they would like in a new development. Cold-calling people can be a waste of time, but sometimes the information received is worth the time and expense. Most developers experiment with different means of gathering information until they hit on procedures that work best for their individual needs.

Traffic Studies

Traffic studies are another form of research than can be important for some developers. For example, when planning to build a mini-mall it would be wise to know how much traffic passes your proposed site on a regular basis. Getting a traffic study done will not tell you everything you want to know, but it will give you an idea of the number of potential customers in a given area.

PRO POINTER

Any type of retail development should benefit from a traffic study. You may be able to find some traffic reports through local agencies, but hiring an independent firm to conduct a study for you will most likely reveal the most relevant data for you to work with.

Traffic studies can be done in different ways. The easiest way is to have a traffic counter installed. This is simply a piece of equipment that counts each vehicle that passes. The downside to this type of study is that you have no idea what types of vehicles are moving through the proposed area. Having a person on hand to watch traffic while the counter records overall activity will give you a better description of the type of traffic that you will be dealing with. For example, if there is construction in the area and there are dozens upon dozens of dump trucks running back and forth, it could skew your numbers. A human spotter could note the high activity of repeat construction traffic.

Proximity Research

Proximity research is an excellent way of matching a piece of property with your proposed buyers. Assuming that you have a general profile of your buyers and their needs and desires, you can do your own proximity research. Assume that you are going to develop a tract of land for a miniature-golf course. What will patrons of the golf course be interested in? Many of the customers will probably be playing golf with their children. Would a nearby pizza restaurant and burger joint be an advantage? Sure it would. How about an ice-cream shop? Yep. Making sure that food and drink are available is a good idea, unless your development is going to offer such services. Then you might prefer a place where such facilities were not so close by.

> **PRO POINTER**
>
> Cutting corners on research is a bad move in most cases for land developers. Your research is your best tool in making wise development decisions. While the cost of professional studies will often run into thousands of dollars, the studies usually pay for themselves many times over.

If you are developing a piece of inner-city land, you might want to make sure that there is good public transportation close by. Many city dwellers rely on public transportation as their only means of mobility. If you are creating an office building in a city, parking could be a major concern. If you don't have enough land to facilitate the parking, how will you handle the overflow? Can you make a deal with a nearby parking garage? These are the types of factors that you need to examine carefully as you pick and choose your properties.

Your marketing team, if you have one, should define your potential buyers. If you are wearing the marketing hat, you have to fulfill that function.

> **PRO POINTER**
>
> Your marketing team, if you have one, should define your potential buyers. If you are wearing the marketing hat, you have to fulfill that function. Someone has to pinpoint the type of buyers that you are going to sell the development to.

Someone has to pinpoint the type of buyers that you are going to sell the development to. Once the target market is known, you can close in on what their needs and desires are likely to be. Once the trigger buttons for what will influence someone to buy into your development are determined, you need to make sure that the needed elements are present. For example, if your marketing team says that you must have a grocery store within one and a half miles of your project, find out if there is such a store. You can pay others to do proximity research, or you can do it yourself. Once you know what you are looking for, there is no reason why you shouldn't be able to handle this phase of your planning personally.

Failed Projects

Most developed areas have some projects that have failed. Some of the projects will have failed before they were ever completed. Others will have been built and then sputtered. You should look into all the projects that have failed in the area where you are thinking of developing land. Looking at the past five years should be sufficient. Find out what failed and why it failed. You can learn a lot from the past, and no developer wants to see failed history repeat itself. For all you know, the land that you are considering may be in a flood or earthquake zone.

How will you get leads about projects that failed? Some of them will be obvious. They will be the ones that have been sitting vacant and deteriorating for the last year or so. Other projects are more difficult to spot. A consultation with an appraiser who works in the area is one fast way of finding out what you want to know. Talking to real-estate brokers is another way to look into failed projects. You could go back through old newspapers and public records, but this would take a lot of time, and talking to appraisers, brokers, and other developers will probably tell you what you need to know.

Growth Status

There are three basic stages within which real estate is rated. There are areas that are in growth patterns. Some areas are stable. Other areas are declining. Another type of area that is not talked about as often is an area that is being revitalized. As you might imagine, you should stay out of declining areas.

Many inexperienced investors buy into stable areas. They can be okay, but they are not likely to produce fast profits and could result in a loss. An area that is stable today could be declining in a year. Growth areas that are established are usually the best buys. Revitalization regions can also be very good, so long as the funding for the projects doesn't run out and sales are as expected.

Making money in real estate can be risky. Buying into new and unknown projects can be especially risky. But it is usually the unknown projects that produce the most profit in the least amount of time. Investors often buy investment properties, such as townhouses, before they have been built. In the business, this is called buying in the dirt. With the right project, an investor can buy from blueprints and sell soon after construction is completed for a tidy profit. This, of course, is assuming that the project is well received.

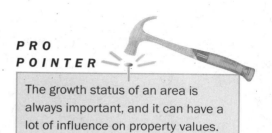

PRO POINTER

The growth status of an area is always important, and it can have a lot of influence on property values.

Developers prosper more quickly when investors are willing to buy in the dirt. Getting contracts for homes that are not yet built makes it possible for developers to borrow against the contracts. Just as investors buy in the dirt in hopes of quick profits, developers sometimes buy on the fly in the same quest. Buying into the unknown is even more risky for developers than it is for investors who are speculating on housing sales. Developers take the first and the biggest risks. You should strive to lower your risk by making sure that the land that you buy to develop is in the best location possible. Put in the time and the money required to reduce your risks.

CHAPTER SIXTEEN

Plans and Specifications

Plans and specifications are a critical element of any development. The plans and specs start with the land and extend into construction. Land planners and engineers usually are the first on the team to create plans and specifications for a development. Architects usually aren't far behind. Of course, not all developments are large enough to warrant the expense of so many professional planners. A small developer may buy stock house plans and have little more than a survey crew involved in general planning. The developer may create most specifications personally.

The degree of complexity that drawings must display can vary greatly. It could take several months to get a general development plan drawn. Even then there would probably be many alterations that would add to the total time for a completed plan. Most jurisdictions require complete plans and specifications before code approval will be issued. Simple plans can work for small developments. A surveyor and some stock house plans might be all that you will need to get a project off the ground. On the other hand, if you are going after a major development, plan on months and months of planning.

Plan Layers

There can be many layers in a development plan. When complete plans and specifications are created for a development that contains a number of components, some of them may include:

- Scaled drawings
- Cross-sectional drawings
- 3-D drawings
- Fill materials
- Landscaping materials
- Road details
- Property lines
- Easements
- Setbacks
- Drainage routes
- Utility locations
- Topography
- Earthwork

Detailed pages in a design plan often provide blowups of small details. Since plans are drawn to scale, small objects can be difficult to see and identify. But when a detail cut is provided, the small object is drawn much larger. A typical scale for precision drawings is one-quarter inch equals one foot. In other words, a line that represents a fence and that is one inch long would be equal to four feet of fencing.

Detail sections on master plans can be drawn as a plan, a profile, a cross section, or any other type of diagram that will be helpful. For example, a small picnic area on a master plan might be little more than a dot on a large piece of paper. To show how the picnic area should look, a detail would be included. Detail pages are used in all sorts of plans. Builders may see detail sheets that show how kitchen cabinets or a particular door might look.

Cover sheets are a part of most plan packages. These sheets offer general information. Cover sheets are usually meant to provide a quick reference to a complete plan package. For example, a cover sheet could include such information as the name of a project, a basic description of the job, a table of contents for what is included in the full package, and so on.

In addition to cover sheets, larger projects may be drawn with an index sheet. An index sheet for a large production plan is similar to the index of this book. Having an index makes it easy to locate specific elements of a plan. Detailed plans for large projects can involve what seems like tons of paper. Having an index to turn quickly to a particular place in the plan is very advantageous.

Types of Plans

How many types of plans will you have to deal with as a developer? It depends largely on the type of projects that you are working with. A very simple development might be drawn on a single sheet of drafting paper. Or you could have stacks of reports containing hundreds of pieces of paper in each report. If you have no other plan, you will have a master plan. In a small development the master plan might only be a page or two long. Complex developments could have master plans created for individual components. For example, you might have one master plan for sewer and water supplies and another for roads. There could be a master plan for construction and for storm drainage.

> **PRO POINTER**
>
> Your overall master plan is the plan that will most likely be submitted for code approval. In some cases code officials may require you to provide master plans for individual elements of a development. Since there are different code officials and different segments of regulations, each phase of the development may have to be approved by individual branches of the code-enforcement system.

The Demolition Plan

A first step in the development process is a plan for altering present conditions. This plan is sometimes called a demolition plan. The name implies tearing something down, like a building, but it can refer to other forms of demolition, such as excavation work. Think of a wooded parcel of land. Which trees will be saved and which ones will be removed? The removal of trees counts as demolition and should be shown on a demolition plan. Underground utilities are another concern when new development is scheduled. It is not in your best interest to demolish underground pipes, conduits, or cables.

Topographical maps are normally used as part of a demolition plan. Any underground utilities must be marked clearly. The person who creates your demo plan has a lot of responsibility. If your contractor digs up a gas main, it could be disastrous. The depth of a demo map varies in detail, but every map should be as complete as possible.

A common consideration in a demo plan is the effect that earthwork will have on adjacent properties and structures. For example, if digging will take place near a sidewalk, fence, or other structure, the structure should be shown on the plan, along with a safe distance for equipment to maintain between construction and the object. Detailed instructions might be provided on angles of excavation to protect adjoining structures, such as parking areas on neighboring land.

The Grade Plan

A grade plan should be drawn to indicate what the finished grading of a project will look like. Rough grading plans should also be included on a grade plan. A typical grade plan will consist of many lines and measurements that will be used to arrive at the desired result when working the earth.

> **PRO POINTER**
> Some developers refer to grading plans as elevation plans. In addition to showing swales, slopes, and finished grades, a grading plan may show trench depths and compaction requirements.

Utility Plans

Utility plans show proposed or existing locations of all utilities on or planned for a site. Most modern utilities are run below ground in developments. It's important that the routing of utilities will not interfere with other development necessities. A detailed utility plan will show the proposed locations of all utility components, such as:

- Valves
- Junction boxes
- Relay boxes
- Transformers
- Buried cables
- Pipes

The Traffic Plan

Traffic is a major concern with large developments. It's common for a separate traffic plan to be drawn for a development. Not all projects require a traffic plan. When a traffic plan is drawn, it may include many details, such as the following:

- Traffic lights
- Guard rails
- Road edges
- Road shoulders
- Driving surfaces
- Pedestrian lanes
- Bike lanes
- Traffic patterns
- Entrances to the main roadways

The Construction Plan

A construction plan will often be drawn to show proposed boundaries for building lots. These plans can show the placement of proposed construction. The plan usually details all setback requirements that must be observed by builders. All sorts of information can show up on a construction plan. Developers who have very strict covenants and restrictions may indicate on a construction plan any number of design requirements for a house built on a particular lot. For example, one lot may be labeled for the construction of a colonial-style home with brown siding and a black roof. Construction plans from developers aren't usually this specific, but they can be. Builders can refer to construction plans to aid in their decision of which lots to buy and build on.

> **PRO POINTER**
> A developer's construction plan is not the same as a builder's plans and specifications. The developer's plan shows an overview of all construction for a development. In fact, the plan is often referred to as a development plan.

Building Plans and Specifications

Building plans and specifications may be a required element in an overall development plan. Developers often call building plans improvement plans. A developer may not be able to gain approval for a development project if the master plan doesn't include an improvement plan. Some developers create their own improvement plans, while others work closely with builders to establish a building plan. Someone has to provide blueprints of structures to be built and complete specifications for all elements of the construction process.

Blueprints for building plans normally show front, rear, and side elevations of the structures to be built. Floor plans are also shown. Framing plans are included as part of a standard set of blueprints. Depending upon the type of project, there can be diagrams that show the routing of plumbing, heating, and electrical systems, as well as any other systems such as fire protection systems. There will normally be detail plans showing roofing, sheathing, siding, and other elements of construction. Exterior plans will show parking areas, driveways, sidewalks, decks, porches, and similar items. Drainage plans for the structure should also be shown in the set of blueprints. A grading plan for the finished grade and landscaping around a building should also be a part of a builder's plan.

> **PRO POINTER**
>
> Specifications for construction should be very detailed. A plan should not call merely for wood siding. Instead, the plan might define cedar siding as the material of choice. The type of roofing material to be used should be itemized within the specifications. It takes time to nail down all the particulars of a construction project, but intricate specifications result in better projects that run more smoothly.

Changes

Changes often occur once a project is under construction. Unforeseen problems can require on-site changes. When these changes are needed, they should be reflected on an amended plan. For example, you might be installing a sewer and run into bedrock that cannot be penetrated. If you turn the sewer to go around the rock, the change in direction should be reflected on the plans filed for your project. Failure to record such changes can create big problems later on if someone has to locate the sewer.

Some changes can be made at a developer's discretion. However, there are times when changes must be approved by code officers or engineers. If you run into a problem on site, you should check with your experts to see if any major consideration should be given to the change before it is authorized. Even small alterations can have a ripple effect. Since most developers are not experts in all aspects of the projects that they create, depending upon expert advice is the best bet.

Cutting Costs

Cutting the costs of development is a goal for nearly every developer. There is nothing wrong with adding to your profit so long as your means of doing so doesn't have a negative effect on your development. Good planners can often find ways to accomplish a goal in a less-expensive manner. It may be up to you to make a judgment call on whether a cost-cutting decision is worthwhile.. But remember that there are times when trying to save a little money can cost you a lot of money.

There can be many times during the planning of a development when you might be given opportunities to reduce development costs. Sometimes the savings can be made without the risk of future problems. Something as simple as using a different type of foundation shrub might save you a lot of money and probably wouldn't offend anyone. You have to weigh the consequences of your actions. If you have doubts, talk to other builders and developers. Ask real-estate brokers for their opinions. Make your decisions carefully and base them on research.

Once your plans are completed, you can bid them out to contractors. This is a critical part of your profit-making procedure. Knowing how to work with contractors and subcontractors can make a huge difference in how much money you make or lose.

CHAPTER SEVENTEEN

Working with Contractors

Getting development bids is a part of the process that can result in more profit than you had hoped for. Experienced developers usually have a good idea of what to expect in terms of contractor costs. There are always gaps to be filled in, but developers with several projects to their credit have a good feel for what costs are likely to run. New developers don't usually have this advantage. But in the end the experienced investors don't always have the best angle. Sometimes rookie developers make the best deal when bargaining for services.

Some developers feel that putting a job out to bid is exciting, and it can be. When you start shopping for prices, you have a chance to increase your profits. If you start into a deal with a certain amount of money budgeted for site work and you shave a few thousand off your estimate, it's money in your pocket. Many see the bidding process as one of the most important parts of land developing.

Independent contractors are frequently used during land development. It is rare to find a developer who has in-house people and equipment for site work. Bids range from surveying to landscaping. Each phase of the development process offers an opportunity for extra income. But the money doesn't come without some effort on the part of the developer. And the tables can be turned so that the bid process results in lost income. If you guesstimated a price for work and the job turns out to cost more, you are on the losing end. It could be said that getting bids is the gambling part of the project. Are you a good poker player? If you are, you just might make your mark on the spreadsheets.

Putting a job out to bid is not as simple as some people believe it is. There are people who think that developers make a few phone calls and get all the prices that they need for a development project. This simply is not the case. Bidding jobs out can be a lot of work, and it is certainly important work. There generally is no shortage of potential contractors to work with, but the quality of some contractors may be suspect. Developers must get more than prices. They must get references and a sense of security about the contractors with whom they plan to work.

PRO POINTER

Estimates are not quotes. Generally speaking, estimates can prove to be worthless, since vendors are not normally required to honor them. Quotes, on the other hand, are firm prices that should be made part of a contract before work is started. Don't confuse estimates with quotes.

Large developers often hire someone to solicit bids for them. It is usually their project manager. Smaller developers generally do the bidding work themselves. I've paid people to collect bids for me, but I've always maintained an active hand in reviewing the bids. I can't recall a time when I authorized any work from anyone I had not met personally. Call it a quirk if you like, but I believe it's important to meet the people who will be working with my company.

We talked much earlier about lining up your experts and giving them examples of the types of services that you would require. Getting bids from some experts is difficult. A lawyer might give you a ballpark figure for certain aspects of a job, such as running a title search, but most attorneys are going to bill their time as they go. Engineers and others are also likely to work on a billable-time basis. Some experts, such as surveyors, may give you fixed prices. It never hurts to ask for committed prices from your team of experts, but don't be surprised if you can't get them. Estimates may be the best that you are able to get. Some contractors will try to get your work on a time-and-material basis. Personally, I would avoid time-and-material deals with contractors. It is in your best interest to have firm prices that you can depend on whenever you can get them. Contractors should be willing to give you firm quotes.

Estimates Are Not Quotes

Estimates are not quotes. Generally speaking, estimates can prove to be worthless, since vendors are not normally required to honor them. Quotes, on the other hand, are firm prices that should be made part of a contract before work is started. Don't confuse estimates with quotes. Entering into a job with nothing more than estimates is risky business. Very few seasoned developers or contractors will do any business based on mere estimates. Quotes are almost always required. There are exceptions, of course, but quotes are the norm. Some special circumstances might warrant a time-and-material basis, but working without firm, committed prices is dangerous.

Estimates can be helpful during planning, but they are so undependable that I would rather not use them at all. There are no teeth in an estimate. I could give you an estimate of $20,000 and wind up billing you $30,000. If the estimate is based on an hourly rate and is not limited to a cutoff level, I could run up a terrific tab for you to foot.

Some contractors use low estimates to lure developers into trouble. It's sad but true that some contractors will stoop to almost any level to get a hook into a developer or general contractor. Creative use of estimates is usually the way that undesirable contractors get hired on a job. If you as a developer insist on quotes, you can avoid the scams associated with estimates.

It would probably be too harsh to say that you should never use estimates. Almost every developer and contractor uses estimates from time to time. But you must be aware that depending on estimates is dangerous. If you are using the estimates as a starting point for estimates of your own, that's fine. But don't count on estimates to see you through a project.

Quotes are the only way to make solid cost projections on a project. Unfortunately, most quotes have time limits that may expire before you can take advantage of them. Locking in quotes for several months can be very difficult. Even if contractors are willing to lock in their labor rates for months into the future, it will be nearly impossible for them to guarantee the cost of materials for a long period of time. This is a problem that all contractors and developers face.

Development projects generally take many, many months to complete. Even homebuilders who can complete a house in 60 to 90 days have trouble locking in the prices for materials. Large builders sometimes have enough clout to get material suppliers to lock in prices. I've been fortunate to develop relationships with my suppliers to lock

> **PRO POINTER**
>
> Quotes are the only way to make solid cost projections on a project. Unfortunately, most quotes have time limits that may expire before you can take advantage of them. Locking in quotes for several months can be very difficult.

in material prices for the full term of construction. Builders who build only a few houses a year don't normally have this luxury.

When material prices can't be locked in until you need them, a quote is not much better than an estimate. If you are planning a large development, suppliers for your contractors and subcontractors may be willing to gamble on a price lock-in. You should at least try to get a lock on all prices. Most developers get the best bids that they can and then factor in an additional amount of money to cover potential increases in material costs. This is a good move, but getting a firm lock is better.

When you begin seeking bids from contractors, you should make it clear that you want quotes, not estimates. The bidding process is tough if you do it right. Don't get into it thinking that your job is easy, simple, or routine. Even if you have worked with contractors before, you have to treat every new project with respect. Let's get down to the art of taking bids.

Locating Contractors

Locating contractors to bid your work can be more difficult than you might think. You can run through the advertisements in your local phone directory, but this approach is a total crap shoot. If you don't already know some reputable contractors, you should start your bidding process by asking for referrals. If you are new to land development, you may not have many contacts in the industry. Don't worry—there are ways to shorten the learning curve.

Assuming that you have chosen your experts, you can ask them to recommend local talent for your development needs. Many developers belong to various clubs. If you are a member of local clubs, you might be able to do some networking that will produce leads for contractors. Builders can also be a good source of leads for subcontractors. If worst comes to worst, you can drive around and find contractors on jobs.

Touring sites in your general work area is a good way to preview contractors before you invite them to bid your work. It's possible to fake references, but it's very hard to disguise work in progress. If you watch a site that is being developed, you can get a good idea of how the contractors on the job work.

> **PRO POINTER**
>
> Many developers ask contractors for references, but the best reference you can get is ongoing work that is being done by the contractors that you are interviewing.

I have often had my field superintendents keep their eyes open for new sites under development. When my superintendents see a project going up, they pay attention to who is doing the work. In fact, written reports from their day logs are saved for when we may need more contractors on our jobs.

Regardless of how you get leads on contractors, you ultimately have to talk with several contractors to get bids on your job. Most developers like to see bids from at least three different contractors. I prefer to get five bids, unless I'm working with contractors whom I've worked with in the past.

PRO POINTER
Before you start asking for bids, you must have all your paperwork in order. By paperwork I mean the plans and specifications for the work that contractors will be bidding on.

Soliciting Bids

Soliciting bids is a job that can be done in different ways. Some developers send bid packages out to all contractors who work within the area of expertise needed. It's common, especially for big jobs, for contractors to be listed in bid sheets that are distributed by companies that maintain subscribers for bid notifications. Most small developers pick up a telephone and call companies to see if they wish to bid work. Once you have your plans and specs ready, make contact with your potential contractors.

You may be surprised at how many contractors will have no interest in bidding your job. This seems crazy. Contractors are in business to work, yet some of them don't want to bid jobs. Why? There are many reasons. Extremely good contractors may be too busy to take on new customers. Some contractors don't have the skills to bid jobs accurately, so they rely on jobs that they can work on an hourly basis. Except for times of major building booms, you can usually get plenty of contractors to offer bids, but it may take more work than you think to find them.

All developers have their own way of soliciting bids. You can do it any way you like, but there are some trade secrets that can make your life easier. For example, let's say that you have five contractors bidding your site work. The prices come in after a week or two and you review them. More time passes as you sort through the bids. You decide on a particular contractor and contact the firm. After meetings and ongoing negotiations, which take considerable time, you find out that the contractor doesn't carry liability insurance and is not bondable. Boom! Your good lead and good bid bites the dust. How much time and money have you lost?

Losing time is bad; losing money is worse. And they usually go hand in hand. How could you have avoided the problem with the contractor? It's simple. You could have required proof of insurance and bonding ability before releasing a bid package to the contractor. Having contractors pass your basic tests before releasing bid packages is not only smart, but it saves a lot of time. Why go through the motions with contractors whom you will not be willing or able to use regardless of their bids?

> **PRO POINTER**
>
> Comparing bids is not as simple as just skimming the bottom line. Even small, simple jobs call for close scrutiny of the bid. If you are the one who is responsible for awarding bids, you must read each bid carefully.

Most builders and developers have certain minimum requirements that subcontractors must meet in order to work for them. Insurance coverage should certainly be one of them. Make a list of what you will require as minimum standards for any contractors before they can work with you. If you are not sure of what you should require, consult your attorney. Prepare the list as a handout for your independent contractors. Include the list of requirements with your bid packages, and make sure that all prospective bidders know that you are firm and nonnegotiable on the issues.

There may be other boilerplate information that you want to include in your bid packages. For example, you might state that you will not allow bidders to substitute materials for those that are specified. Material substitution is a common ploy used by some contractors who want to make their bids look attractive. You can't afford to accept this type of activity. Make it very clear that all specifications must be adhered to exactly. In the event that a substitution can't be avoided, insist that the bidding contractors provide a detailed report on the material substitution being offered. This will give you and your experts a chance to check out the substituted material for quality and value.

Comparing Bids

Comparing bids is not as simple as just skimming the bottom line. Even small, simple jobs call for close scrutiny of the bid. If you are the one who is responsible for awarding bids, you must read each bid carefully. You can bet that there will be differences between the various bids, and I don't mean just the price. Contractors bidding a job usually make some alterations to the provided bid package. In theory, reviewing the bids should be as simple as looking at prices, but it seldom works out that way.

If you have prepared a detailed bid package that leaves nothing to the imagination, you might get clean, easy-to-compare bids. Otherwise, you will have to sift through the bids and compare them closely. Some developers use bid packages in which a contractor agrees to price all the work contained in the package, fills in a blank with a price for all labor and material, and then signs the bid sheet. This is an ideal situation, but it is rare. It is much more common for contractors to offer you a letter as a bid or perhaps a quote on some type of form from their company. Receiving bids of this type is more risky for you. These are the bids that you have to go over with a magnifying glass.

I've received some very creative bids in my time. It's amazing how many contractors and suppliers fail to include certain details in their bids. To keep your prices dependable, you have to make sure that the bids are equal in all respects. You know, it's the old apples-to-apples thing. If you have three bids that are not identical in their description, they are nearly worthless. Prices for lemons, apples, and oranges can't be compared equally. You have to get bids that can be compared properly. Contractors don't always want you to have such contracts, so insist on it. If they want the work badly enough, they will play by your rules.

Some subcontractors have to be convinced to bid jobs in a way that is acceptable to you. There are contractors who only bid jobs their way. Forget these people. If you can't get them to bid a job to your specifications, just imagine how difficult they will be to work with during the course of your project. You're the boss, and you have the right to request bids to your standards. Don't settle for less.

Refining Bids

Refining bids may be necessary. Assume that you have encountered bids in which some materials had to be substituted. It's unlikely that all of your contractors substituted with the same alternate materials. You may have to refine the bids to make them equal. This could be as simple as calling the contractors and having all of them agree to bid with identical substitutions. You may decide to take this step by mail or e-mail. Face-to-face meetings might be the best way to distill comparable bids. The way in which you do it is not nearly as important as getting it done.

There may be other reasons for refining bids. You might find that you wish to make changes to your plans and specifications after a bid package goes out. Developers hope that this will not happen, but it sometimes does. The development business is full of surprises and changes. Just when you think that you have a firm plan, it may become evident that something was missed or must be changed. If this happens, you should get in touch with your contractors and refine any bids that you have requested or received. A written change order should be issued so that there can be no dispute as to what the changes are.

Putting up with the paperwork of a development project is more than some developers wish to deal with. The volume of paperwork can be extreme, but it is necessary. If you don't have the patience or organizational skills to maintain good paper trails and files, hire someone to do it for you. Don't attempt to eliminate the paperwork. Sooner or later, developers without documentation generally wind up in trouble.

CHAPTER EIGHTEEN

Projecting Profit Potential

Projecting profit potential is an essential part of successful land developing. No experienced developer will launch into a development deal without accurate sales projections. Some developers create their own sales projections, while others hire consultants to do the projections for them. Many deals never get off the ground when sales projections will not support them. It's good business to make sure that there is adequate demand for what you plan to develop. You also have to make sure that your potential stable of buyers can afford to pay the price of admission. Having buyers lined up around a street corner is of little value if they don't have enough money or credit to buy what you are trying to sell.

The preparation of sales projections is a process that can follow proven paths, yet much of the process is supposition. You can't be sure of what will happen until you attempt to make it happen. One problem with land developing is the time that it can take to turn out a salable product. What starts as an excellent idea can turn sour quickly. You could be caught in a quagmire of financial problems. It has happened to me and many other experienced developers, so it could happen to you.

Making Sales Projections

There are various methods available to you for projecting sales, some of which are included below:

- The easiest way to get sales information is to hire a consulting firm to perform the sales research for you.
- The downside to hiring consultants is the cost and the lack of personal involvement.
- Many developers prefer to be involved in their own research.
- Historical data is one of the best ways to determine what has happened with previous development efforts.
- You could have a canvassing firm call people who fit the demographic profile for your project in an effort to get detailed information on what they would be interested in buying.
- Talking with builders can also be a good way to project sales.
- The most dependable means of sales projections is usually the comparable-sales method using historical data.
- Developers and builders often have gut feelings that are well worth following.

Personally, I've been successful in using historical data as my foundation and my gut feelings for the window dressing.

Local Trends

Many housing projects take years to develop and sell out. A lot can happen with trends over the course of years. I've seen developers fall into the trends trap. I have seen earth-sheltered houses fail, dome houses bite the dust, and passive-solar homes go down in flames. All three types of housing were trendy for a while, but the trends didn't last when the developers were trying to cash in on them. I watched as development after development got off to a good start and then fizzled out. While this was happening, I was developing subdivisions of contemporary homes and colonial homes, which all proved successful. My house designs were not radical or trendy, but they stood the test of time.

PRO POINTER

Local trends can stimulate developers to follow a growing pattern of activity, but the trends may die out before a developer can complete a project. This can be a financial killer.

You may not be involved with the construction of homes. Maybe your market is homebuilders and you are providing only building lots. Even straight land-development projects can fall into the trendy traps. If you get too cute, you could lose money. Some examples of mistakes could include the following:

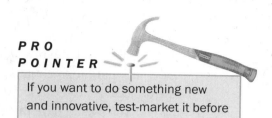

PRO POINTER

If you want to do something new and innovative, test-market it before you commit to it.

- Cutting lots into strange shapes
- Creating high-definition landmarks
- Eliminating expected amenities
- Adding too many amenities

Developers who stick to the basics usually have the longest history as successful developers. If you want to do something new and innovative, test-market it before you commit to it. Once you set the tone of a development, it is hard to change it. And the wrong public perception can lead to costly losses in sales.

Working with a Team

Working with a team of real-estate professionals is an excellent way to develop your own sales projections. Real-estate brokers will often work with developers in the creation of sales projections. In most cases the brokers don't charge for their services. They do the work in the hope of getting the listings on property in the development. Real-estate appraisers can also provide deep insights into probable sales. Unlike brokers, most appraisers will expect to be paid for their services when they are rendered. Even at high hourly rates, the information derived from appraisers might be one of the best bargains you will find in the development business.

Some developers are members of multiple-listing services (MLS). Real-estate brokerages nearly always subscribe to multiple-listing services. If you are not able to access information from a multiple-listing service on your own, you can gain access to it through a real-estate broker who is a member of the system. The information available through MLS is very valuable when developing sales projections. Details vary, but most of the systems provide all data needed to determine comparable sales prices.

You could develop your own team of sales professionals to work with while creating sales projections. For example, you might hire an appraiser, work with several brokers, and track the results of a local MLS. You might also talk to local builders to get their input on potential sales. If you choose to hire a consulting firm to produce sale projections, you can factor in their findings with those of the rest of your team. In the end, you should have the makings for an accurate projection.

Land Only

The value of building lots is often established based on the value of completed housing packages. In my area, builders and banks usually consider the value of land to be about 20 percent of a total housing package. In other words, a complete house on a building lot might sell for $200,000. If this were the case, the value of the land might be estimated at $40,000. There are certainly times when land is worth more or less than the rule-of-thumb 20 percent.

How will you establish the sales value of your land? The safest route is to have a complete appraisal done. This is expensive, but you can count on it beyond any other type of sales projection. The appraisal method is so simple and involves so little time or effort that it seems that more developers would use it. But the cost can be prohibitive. If the appraiser used for the job is on the approved lists of local lenders, you can rest comfortably knowing that you are playing with real numbers. Any other approach leaves more room for error.

> **PRO POINTER**
> If you are developing land only, you may be able to get by with a lot less research than a developer who is also a builder can. However, don't assume automatically that just because you will not be building houses that you don't have to know about housing prices.

The next best method of putting a price on your property is, in my opinion, the comparable-sales method. Here access to MLS data is invaluable. By looking at closed sales in the MLS you can see what people have really been paying for land like yours. The comparables, or comps, as brokers call them, should be near your land, and the sales should be less than six months old. If you find three developments in an area near where your lots will be and the lots in those developments are selling for prices between $32,000 and $35,000, a price of $40,000 could be a stretch for your project. If you expect to get that much more, you are going to have to make it enticing for buyers. But convincing buyers to pay the price may not be enough. If the land is being financed, the appraised value will be what counts when it comes to loan approval. Even if buyers are willing to pay $40,000 for lots, their lenders might loan against the lots as if they are worth only $35,000 or less.

> **PRO POINTER**
> Before you gamble on getting more money for your development by adding new twists to your design, make sure that the improvements will pay for themselves.

Before you gamble on getting more money for your development by adding new twists to your design, make sure that the improvements will pay for themselves. You can only do this by working with an appraiser. It's good business to expect a return on your investment. The return doesn't have to be in the form of cash. For example, an improvement that makes a development sell faster but for no more money can be justified. However, not all improvements are worth their expense. Investing money and getting back less than what you put in is an excellent way to go out of business in a hurry. Assuming that you want to stick around as a successful developer, you have to make investments that pay for themselves.

Land and Improvements

If you are developing to sell land and improvements, you have more work to do. Not only will you have to peg a price for the land; you will also have to pin down a value for the improvements. In the case of houses this shouldn't be difficult. But if you are developing something that is a bit rarer, the prospect of picking the right prices can be daunting. Let's say, for example, that you are building a development of duplexes in a section of town where there are no duplexes. Your idea might be terrific and sales may be brisk, but how will you nail down a price? Since there are no other duplexes in the area, you will have no comparable sales data to work with. It can be tough to pick a price and predict sales in such a case.

Will buyers flock to your duplex development? How can you know? Demographic studies and local research are about all that you can go on in this type of situation. Most of your projections will be difficult to prove. When you get into a deal that is unique enough to be without historical data on comparable sales, you are flying solo. Without closed sales to compare your development to, it is nearly impossible to make a comfortable projection on price and sales.

Should you avoid developments that can't be supported by comparable sales? If you are a conservative person who doesn't like to gamble, you should. However, if you are looking for the next major trend or a windfall of money, going where no one has gone before might be the only way to realize your goals. It is risky. You don't know what to expect. Sometimes you just have to go with your gut feelings and hope for the best. The biggest money is often made from the most dangerous deals. But a deal doesn't have to be dangerous to be profitable.

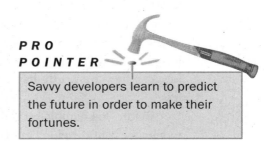

PRO POINTER
Savvy developers learn to predict the future in order to make their fortunes.

Beware of Hype

Beware of the hype that some real-estate professionals will heap upon you. In addition to being a builder and developer, I'm also a designated real-estate broker and own a brokerage. I can tell you that some listing agents will push the envelope considerably to secure a listing. Personally, I don't approve of too much hype, but I see it happen from both sides of the table. In the real-estate industry the hype is often identified as "puffing." There can be a thin line between puffing and misrepresentation.

If you are working on a major development, there can be a lot of money to be made by the listing broker. Plenty of brokers will want your business. Most of them will be honest with you, but you might run into a few who will embellish the reality of your situation, and this is putting it nicely. It's common for brokers who are trying to secure listings to offer every type of promise imaginable to a seller.

Be careful if you go to brokers for help in creating sales projections. Some brokers want high sales prices for two main reasons. The first reason is that the commission earned from a sale is higher when the sales price is higher. The second reason is that sellers are more likely to list their properties with brokerages that suggest the highest market value. None of this means anything if you don't get sales. Listing agents get a brownie point with their sales manager, and a brokerage loses money advertising overpriced property, but you—the seller—will not see sales in quantity or quality.

Real-estate brokers can be very helpful in projecting sales prices and sales activity. Check their data to make sure that the numbers that you are working with are reasonable. Don't get caught up in the hype of talk that is not substantiated with facts. Base your sales projections on proved facts.

Anyway you cut it, sales projections are little more than guesstimates. Some are better than others, but none of them is a sure bet. Even the most formal, best-researched projections are unproven. Until you are knee-deep in the development process, you will not know how accurate your projections are. A sudden shift in interest rates can kill your projections. Other elements, such as tax-law changes, the closing of military bases, and other outside forces can ruin the best plans for a development. The best insurance is to get in and out fast before there are significant changes in the market. Make the best projections that you can, and then run with the wind to make your deal happen while the conditions are right.

CHAPTER NINETEEN

Zoning Considerations

Zoning has much to do with what developers are allowed to create. The laws, rules, and regulations for zoning can vary greatly from town to town and city to city. You can't go by a state-by-state formula when it comes to zoning laws. Every organized community is likely to have its own zoning regulations. And zoning laws can be changed fairly quickly, so don't rely on old zoning decisions. Every piece of land can fall under different zoning ordinances.

Zoning regulations can be very complicated. Even experienced real-estate attorneys can have trouble interpreting the laws. Trying to make sense on your own of all the zoning issues that you will deal with as a developer would be crazy. You will need a good lawyer to work with you to cut through all the red tape associated with zoning. However, there is a lot about zoning that you can understand. These issues are what we will cover here. But remember to consult your attorney for clear interpretations of zoning before you make a buying decision on a parcel of land.

Zoning Maps

Zoning maps are good starting points for checking the status of land. One purpose of zoning is to prevent conflicts in land use. It is the zoning laws that balance a community in what is believed to be the best means possible. Zoning maps are drawn to show specific zoning regions. For example, you might see that one part of town is zoned for retail use, while another section is zoned for industrial use and another is zoned for residential use. It is common for zoning maps to be altered periodically. Requests for zoning changes are sometimes approved. When they are, some reference to the changes must be recorded. Eventually, the zoning maps reflect the changes.

> **PRO POINTER**
> Consult your attorney for clear interpretations of zoning before you make a buying decision on a parcel of land.

If you are researching a particular piece of land, you can look on a zoning map to see what the existing zoning is. When the established zoning is compatible with the type of project that you want to develop, you have it made. It is a simple matter to move forward when you don't need a zoning change to do it. If, however, you find that the land that you are considering is not zoned for a use that will allow your type of development, you have some work to do.

Some types of zoning changes can be reasonably simple to obtain, but others are very difficult. It is certainly easier and less expensive to develop a piece of land that is already zoned for the proposed use.

I have had my share of dealings with zoning officials and boards of appeals. The experiences have not been fun. Fortunately, I can't remember a time when I did not prevail. But this is not to say that winning was easy or that it came without a fight. Some of my zoning battles have been very expensive. They have also been quite time-consuming. If a project doesn't have enough potential, fighting the system isn't worthwhile. Before you engage in a major zoning battle, make sure that the time and money that you will invest in it will be rewarded in the end.

Zoning boards usually have a lot of latitude in how they handle their regulations. A landowner might obtain a variance with relative ease. When a variance is approved, it is usually in the form of a minor variation from existing zoning requirements. As an example, a variance might be issued for a garage to be built two feet closer to a property sideline than what present zoning requirements call for. It would be

> **PRO POINTER**
> If a change in zoning will be required for a project, you could be looking at many months of legal maneuvering and a lot of out-of-pocket cash expenses. Your choices are to pass on the property and look for another parcel that will be zoned for your needs or to file for a change in zoning.

common in such a case for a developer to be required to talk with neighbors who might be affected by the variance and to gain permission from the neighbors before the variance is issued.

Zoning maps show you what existing conditions are. The maps do not indicate that the land use might not be changed. In some cases, the zoning maps might tip you off to great opportunities. Finding land where zoning has been changed for a higher and better use can result in much higher profits for a developer.

> **PRO POINTER**
> If you can buy land that the owner believes is zoned as residential and then use it for commercial purposes, the value of the land should soar. This is not all that uncommon.

Many residential areas slowly change over to other types of land use. This usually happens as an area grows. Car dealers, fast-food restaurants, hardware stores, and all sorts of other nonresidential uses move into an area. As this happens, the houses are sometimes converted to new uses, or they may be removed to allow a higher use of the land. People who live in the houses sometimes sell their property for huge profits. But some sellers are not aware of how much more their homes are worth as the zoning laws and land uses have changed. Astute developers scour zoning changes in search of rare opportunities. Zoning maps may not show the changes soon enough. Dig deeply when you are researching permitted land use. Your research can keep you out of trouble and may make you much wealthier.

Cumulative Zoning

Cumulative zoning is a type of zoning that often allows for changes in regulations. Developers tend to like this type of zoning, since it can allow a great deal of freedom. Communities, however, sometimes suffer from cumulative zoning. The purpose of zoning is to manage land use and to maintain certain separations. Some strange results can occur with cumulative zoning. For example, you might find housing developments mixed in with commercial projects.

The value of land is often in direct relation to the zoning laws. Obviously, a piece of land that can have a shopping mall built on it should be worth more money than the same piece of land if it is limited to single-family housing. Striving for the highest and best use is a good goal, but it is one that some communities cannot afford to enforce. The result is often a mixed-use community, which allows different types of land use in the same area.

Why would a community vote for mixed use? Deriving tax dollars from landowners might be one reason. In some areas strict zoning keeps residential areas so far from places of employment that traffic becomes a problem. If people have to move into fringe areas and commute to work, the traffic flow can be more than the roads in a community can handle. There can be any number of reasons for cumulative zoning.

Floating Zones

Floating zones are usually districts that are not mapped for specific uses. Communities may use floating zones to give themselves flexibility in applying their regulations. Since floating zones are often unmapped, they can be difficult to pin down during preliminary research. A floating zone may be used as a means to move into transitional zoning. Planned unit developments (see below) often come into floating zones.

> **PRO POINTER**
> Floating zones are common areas for PUDs. Many communities encourage the development of PUDs. But there is no guarantee that a project will be approved, and this is a risk.

Transitional Zoning

What is transitional zoning? It is a type of zoning that starts as one type of land use and gradually changes as distance increases. This type of zoning usually works well. It can facilitate a number of land uses without hurting the look of a city. For example, the strictest end of the zoning might contain heavy commercial properties. As a development sprawls out, the zoning could change to light commercial, then to office space, and eventually to residential use. You might even find regulations that require office space to be developed to resemble residential architecture.

There is a town in Maine where low-impact features are required of many types of business property. For example, a major fast-food chain is housed in what could appear to be a large farmhouse. Many of the restaurants and businesses are required to be housed in buildings that are compatible with the residential area. This can make it confusing for people who are looking for the business image that they have come to be so familiar with, but it does enhance the quaint appearance of the town.

Planned Unit Developments

Planned unit developments (PUDs) are quite common. This type of development may house everything from single-family homes to commercial retail stores. It is common for a PUD to be a stand-alone community. The community may include all forms of service businesses, medical facilities of some sort, and a variety of housing types. Due to the mixed use in a PUD, careful planning is needed on the developer's part.

Floating zones are common areas for PUDs. Many communities encourage the development of PUDs. But there is no guarantee that a project will be approved, and this is a risk. The money spent designing a PUD for application approval can run into thousands of dollars. This is money that will be wasted if the project is not approved. Gaining approval for a PUD can mean major profits for developers, but the venture capital invested to get the approval can prove to be a substantial risk.

When a PUD is designed, it can contain many types of buildings and land uses. Most commercial uses are required to be neighborhood-serving establishments. Generally, PUDs are required to provide open space, recreational facilities, and other amenities. It is common for some provision to be made for local fire prevention and protection. Schools may also be a part of a PUD. The expense of developing a PUD runs high. But a PUD that is well planned should sell well. Few first-time developers start with PUDs.

When a PUD is designed, sections of the land are set aside for different types of uses. These sections are identified on a site plan. For example, one section may be given a rating of R-1, which could mean low-density use. A section with a rating of R-20 could be set aside for ultra-high density, and there could be other ratings between the two extremes.

Cluster Housing

Cluster housing is popular in large urban areas. Some communities approve cluster developments in an attempt to preserve land around historic sites or even for valuable farmland.

> **PRO POINTER**
> The general concept behind cluster zoning is a balance of smaller house lots mixed with larger open areas.

Developers who can gain cluster approval can turn a small parcel of land into a valuable housing project. There usually is not as much flexibility with cluster zoning as there is with planned unit developments.

Zoning density is usually maintained in a cluster development. However, the lot size and setback requirements are often forgiven. There is not as much freedom with a cluster development as you might think. Even though you can reduce lot size, you may not be able to increase the number of housing units. Remember, the goal of clusters is to provide more open space per dwelling unit. Many developers think of clusters as a way of squeezing a lot of housing into a small space. This is sometimes possible, but don't expect it to be the rule; it is the exception in terms of density.

Exclusionary Zoning

Exclusionary zoning is intended to prohibit specific types of development. This type of zoning is used to keep a certain type of development out of a region where general zoning would allow the use. Take, for example, a region where clubs are allowed. There might be exclusionary zoning to prevent the creation of a gun club that includes a shooting range. The noise from the firing range might be the reason for excluding the land use. Noise, odor, and pollution are common reasons why exclusionary zoning is used.

> **PRO POINTER**
> Exclusionary zoning is intended to prohibit specific types of development. This type of zoning is used to keep a certain type of development out of a region where general zoning would allow the use.

Gun clubs might not be the only target of exclusionary zoning. This type of zoning can limit any type of land use, at least in theory. For example, the zoning might exclude the construction of apartments or mobile homes. Minimum house sizes might be required. This is not uncommon for a developer to require in covenants and restrictions, but it is odd to have a zoning regulation of this type. In fact, exclusionary zoning often comes under fire as being a form of discrimination.

Inclusionary Zoning

Inclusionary zoning is designed to promote the development of both low- and moderate-income housing. To do this, communities offer more flexibility in their zoning regulations. For example, a developer may be allowed to increase the density of a development in exchange for controlled-price dwellings. This can sound good on paper, but it may not work so well in reality. Not all land parcels can be maximized with the type of housing that you may want. For example, you may have to include townhouse designs to achieve the higher density.

Other Types of Zoning

There are other types of zoning. In fact, you may run into any number of zoning situations as you move from one jurisdiction to another. Zoning laws are similar to building and plumbing codes. They are usually offered in a generic form that is adopted and adapted by local governing bodies. The adaptation can be extreme. You and your attorney will have to read all the fine print to stay out of trouble. Don't make any assumptions. Zoning is a major part of the development process, and it is a topic that you will have to become acquainted with.

CHAPTER TWENTY

Closing Deals

Closing deals can be simple and require little of your time, or it can be a challenge to your patience unlike any other. In theory, getting from an accepted purchase agreement to ownership should be a routine matter that merely takes time. Most closings take between 30 to 60 days. Once you have a fully executed contract to buy land, your dealings should be in the hands of others and off your plate, but don't count on it. It is not uncommon for major problems to arise somewhere along the way during the closing process.

Given normal conditions, your attorney and other professionals will be running the show to get your purchase settled. There can be a great number of people involved in the overall process. Some of the players may include your lawyer, the seller's lawyer, real-estate brokers, surveyors, title companies, insurance companies, property inspectors, appraisers—and the list goes on. Developers hope to be hammering out details for development while papers are being shuffled for closing. There are many deals that do close smoothly, but there are some that don't.

By the time you reach the point of being ready to close a deal, you may have invested a substantial amount of money on a deal that you don't yet own the land for. Engineering reports, soil tests, and similar work done to give you the green light to buy a property are all paid for out of your pocket. If the sale doesn't close, you have lost the money. The seller will not reimburse you, and few experts are going to return their fees if you can't close your deal. It's possible to lose thousands of dollars before you ever own a piece of land.

How can you protect yourself during closing procedures? Having a good lawyer is a start, but even that doesn't guarantee success. There is so much that can go wrong that you can't protect yourself from everything. Much of what might happen is hidden until the closing process begins. You can, however, do some work ahead of time to limit your risks.

If you are not an expert in dealing with real estate, you will be limited as to what you can do personally. But you can still maintain an active role and keep your closing on track. I'm fortunate to be a designated real-estate broker with many years of experience. I know the ropes, so to speak, so I can do a lot of my own work. You may not have this advantage, but you can still do more than most developers do to ensure a timely closing.

The Closing Process

The closing process begins once a buyer and seller agree to a transaction. There are usually many contingencies that must be removed from a contract for purchase. Time is needed to clear the contingencies. One of the early steps in the closing process for most buyers is the application for financing. If you have your loan package prepared, this step only takes an hour or so. Once a relationship is established with a lender, the rest of the closing process moves into gear.

Lenders will want current appraisals on the property being bought. The appraisal process when buying a home might take less than a day, but the same process for a development project could take several days or more. It is during this process that information is collected to place a value on the existing property and the property as it will be after proposed development plans take place.

Someone, usually a lawyer or a representative from a title company, has to perform a title search. This aspect of the closing process can go quickly, but delays may occur with some properties. A typical title search involves tracking all activity involving the title to the land over the past several decades. The search could go back to ownership for the last 40 years or more.

Current surveys are often a requirement of the closing process. Depending on circumstances, the fieldwork for a survey may take a few hours or several days. The time and expense are related to how difficult it is to pin down property boundaries. Some properties have iron stakes and benchmarks that are easy to locate. Other properties may be much more difficult to survey, such as when a property is being cut from a larger parcel or when there have been no surveys done in recent years.

Employment verifications are a part of the closing process, and they are handled by the lender. Many developers are self-employed. For most, this means providing tax returns for at least the last two years. The verification process includes checking bank accounts, income, credit history, debt service, and other factors that may be deemed important to a repayment plan.

In simple form, closings are not complicated or difficult. There can, however, be much more to the closing process than what we have discussed. For example, you may have to apply for and receive permits for development before you can close your loan. It could be necessary for you to provide detailed documentation on development costs before closing, and this might mean obtaining many written quotes from contractors. You probably won't conduct your own title search or perform your own appraisal, but there is work for you to do.

The Appraisal Process

The appraisal process starts early during closing procedures and is very important. If the land that you are trying to finance doesn't appraise well, you will not be able to borrow as much money against it. You can't do your own appraisal, but you can work with the appraiser. All good real-estate brokers know the value of working with appraisers. If you want to make sure that you get a high appraisal, you may have to do some research for your appraiser. The appraiser might be appointed by the lender or may be someone whom you have hired that the lender will accept. You can have an impact on the appraised value of your property. Here are some of the things that appraisers look for and what you can do:

- Appraisers look for all sorts of information when putting a price on a piece of land.
- If you have access to information that may not be in a current MLS, you should make it available to your appraiser.
- If you know of a private sale that closed within the last six months that would be comparable to the property that you are having appraised, the information could be very valuable to you. Give the information to the appraiser.
- Depending upon the current market conditions, an appraiser may have 10 to 12 potential comparable properties to choose from for appraisal purposes. The appraiser may use only three pieces of property as comparables. The three chosen can have much to do with the value placed on your land.

- Appraisers should look for comparable properties that are the best match with your land.

In many ways, the appraisal process may be the most critical part of a closing. If your land doesn't come in with a value high enough, you will either have to walk away from it or put a larger portion of money on the table as a down payment.

The Title Search

The title search of a property is often done by a developer's attorney. Sometimes the search is done by a title company or someone else. Searching a title is not particularly difficult, but it is a critical part of long-term success. I often do my own title searches before making an offer to buy land. Then I have my attorney do a title search before closing on a property. While I know how to do a search, it suits my style to have the backup of an attorney. And most lenders want a title search to be done by someone in a legal profession.

> **PRO POINTER**
>
> A clouded title is one where there are problems or potential problems that might make the transfer difficult.

What do title searches turn up? In most cases they are just part of the closing process and don't prove to be monumental. But there are times when a title search kills a deal. This is why I like to search a title before I make preliminary investments prior to a purchase. Title searches are intended to find anything that might cloud a title. Some title defects can be insured to make a property fairly safe to buy, but other problems are almost insurmountable.

The types of problems found during a title search can be numerous. Some examples of these problems may include the following:

- Unpaid taxes
- Tax liens
- Ownership disputes
- Illegal subdivisions
- Nonconforming use
- Mechanic liens
- Materialman liens

Closing Deals

Surveys

Surveys are almost always a part of the closing process. It's possible that a seller will have a current survey available that will relieve the need for a new survey during the closing process. If this is the case, it's fine. However, if there is not a recent survey available, you should insist on one. Land is often what it seems to be, but sometimes it is not. You can't depend on boundary markers provided by sellers. Many owners lose sight of where exact boundaries are. Unless you are able to find verifiable survey markers, you should question the boundaries of a property. During the closing process, a survey can turn up undesirable information. If a section of land is not as it has been reported to be, a closing might be stopped.

> **PRO POINTER**
> If there is not a recent survey available, you should insist on one.

Income Verifications

Income verifications are another part of the closing process. There shouldn't be any surprises in this procedure. As long as you report your income as it really is, there shouldn't be any problem. However, you may have to provide more verification than what you are expecting, and this might be a problem. If you are self-employed, you can expect to provide a minimum of the last two years' tax returns. Generally, a lender will average the earnings of the two tax returns to arrive at your income. You may, however, have to provide more detailed information.

> **PRO POINTER**
> If your purchase contract has a "time is of the essence" clause in it, you could lose your closing by not having your documentation handy. When you make a loan application, make sure of exactly what will be required of you and comply as quickly as possible.

Credit Reports

Credit reports are a part of the closing process. Typically, a lender will pull a credit report on you before sending a loan to the closing process. But a situation can change between the time that you start a closing and when you finish it. Just as you will want a "fresh" title search, a lender is likely to want a fresh credit report before closing. If anything happens during the closing process to taint your credit rating, you could lose your deal.

Credit problems can come from a variety of reasons. Usually they are associated with situations that people who suffer from them are aware of, but this is not always the case. There are times when dark spots can appear on your credit report without your knowledge. Sometimes reporting agencies make mistakes.

Tracking

A typical real-estate closing involves a lot of people. It would seem logical to assume that these professionals will do their jobs independently. Don't count on it. It's almost scary how unorganized some closing procedures are. Having been in the real-estate business for decades, I've dealt with too many closings to count. The number of times that one hand doesn't know what the other one is doing is simply incredible at times. The level of poor communication between players in the closing process can be extremely frustrating. This is where a broker's intervention or a developer's supervision can make the difference between a timely closing and a busted deal.

The closing process is not usually overseen by a single supervisor. All the people involved in the process are working independently. Lenders are the closest to functioning as an overall supervisor, but they don't always keep as tight a rope on the closing process as some developers would like. If you have a good broker representing you, the broker will take care of your closing for you. Developers who are not working with brokers, or who don't have enough confidence in the brokers with whom they are working, should track their own closing process. Failure to track a closing can result in a blown deal and lost money.

Keeping tabs on your closing is not a major task. It will, however, require you to make some phone calls to make sure that everyone is on schedule. Assuming that you will be doing your own supervision, you will need the names, phone numbers, and maybe the e-mail addresses of all the players on your team. Your goal will be to track all elements of your closing. This will basically be a weekly task that could last for up to eight weeks and maybe more.

Essentially, you should check in with everyone involved in your closing. If a survey is supposed to be delivered by a certain date, call on the following day to make sure it arrived. If it did not, call the survey firm to establish the status of the survey. This process will apply to appraisers, attorneys, lenders, inspectors, and so forth. You should have a complete list of everyone working on your closing. The list should be set into a schedule so that you can know if every aspect of the closing is on time. You can lay out your table of players and tasks on a computer or on a sheet of ruled paper. The important thing is to have every category, every player, and every date in place.

Once you have your schedule, post it in a place where you will see it regularly. It's easy to become distracted by the many hats you wear as a developer. Putting your schedule in plain view can be quite helpful. Record anticipated dates. Insert space to note when tasks are completed. If there is a problem, and there will probably be some, make more notes of what you need to do to solve it. Stay on top of your list. Work it daily if need be.

Your active participation in the closing process can be the difference between a successful closing and a lost opportunity. Developers who rely too heavily on others are often disappointed in their success ratios. You can swing the odds in your favor by keeping an active hand in the closing process. Remember, without a successful closing you can't do much development work. One key to the success of developers is building each block of the development step by step, and the closing process is one of those building blocks.

CHAPTER TWENTY-ONE

Supervising Your Site

Supervising your site during development is essential in maintaining a quality project. Even if you hire an on-site manager, you should stay involved in the supervision. It is not necessary to do it all yourself, but you should keep yourself active in the supervision of the site.

How much time will you have to spend in a supervisory role? That depends on how well you build your development team. If you do your homework, get the best people you can find, and hire a competent project manager, your role in supervision can be minor. When things don't work out so well in the planning stages, you will have much more of a demand on your time.

The requirements of supervising a site vary with the size and type of project being developed. Even small developments require proper management, but the larger the tract of land is, the more there is to pay close attention to.

Going Solo

If you choose to go solo on your site supervision, you will need a full understanding of the principles and practices involved with creating a successful development. This usually comes with experience. While much can be learned through traditional education, there is nothing quite like having your boots in the dirt to get you ready to develop your own site.

Should you assume full responsibility for your site? Probably not, at least not unless you have other successful developments under your belt. Developers become knowledgeable in various ways. Some go to college and study the development process. Many work for developers and learn in the field. Some educate themselves, take the plunge, and survive it.

Personally, I combined multiple methods in becoming viable as a land developer. I was very active in both real estate and construction prior to venturing into land development. My diverse background around the edges of land development enabled me to earn while I learned. Additionally, I immersed myself in self-study. Before taking on large developments on my own, I partnered with an experienced developer on a few deals. The experience was invaluable. That's the way I did it. How will you do it?

Over 30 years of activity, I have become a fairly astute developer. I depend on myself for many of my decisions, but I am always consulting experts in specific fields. Even though I could run my own projects individually in a supervisory role, I prefer to have a project manager on site for day-to-day duties. For me, this is a wise use of time. I suggest that you factor in the expense of an experienced project manager when assessing your profit potential.

Going it alone can be a long and troublesome route. Even with adequate experience, it is a heavy burden to bear. There will be sleepless nights, countless questions, and confusion. Having a site manager to talk over the details with during the development process is very beneficial.

What Qualities Should I Look For?

What qualities should I look for in a project manager? The skills needed vary from project to project. For example, a residential development can be very different from a commercial development. When you are creating a sustainable environment, there are some common traits to look for in prospective managers. The list below outlines some of the factors for you to consider when choosing a project manager.

- How many successful projects has the individual managed as a senior manager?
- What percentage of the person's projects have failed to meet expectations?
- What type of experience does the candidate possess?
- Is the project manager more experienced with commercial or residential developments?

- Has the prospect ever managed a mixed-use project?
- How much knowledge of sustainable procedures does the individual possess?
- Will the individual pass a background investigation?
- Has the potential manager worked for the same developer on multiple projects?
- Have there been code or safety violations on previous projects managed by the individual?
- Have projects come in on time and on budget?

The list above gives you a good idea of the types of questions to ask of a potential project manager. Remember that the project manager you choose will have your money, your project, and your future at stake.

What Will You Expect?

What will you expect from your manager? Will the manager be responsible for financial reports? Is the primary function of the job to schedule and supervise subcontractors? Define the job description. Interview prospects carefully. Make sure you are getting what you are paying for. To clarify, consider the list of potential duties below for a project manager:

- Arrange, coordinate, and supervise engineers.
- Be available on site to meet appraisers, bankers, insurance inspectors, safety inspectors, and so forth.
- Schedule and supervise subcontractors.
- Arrange deliveries.
- Manage refuse removal.
- Coordinate site sanitation requirements.
- Track scheduling requirements and deadlines.
- Confirm code inspections.
- Compile financial receipts and data.
- Prepare management reports.
- Daily to weekly meetings with you.

The Decision Is Yours

The decision to hire a project manager is yours. Not everyone wants an employee between them and the management of their development. It is a personal choice. I like having someone on the site when work is in progress and I cannot always be there personally. If you decide to delegate some of the supervisory duties, do it with diligence. It is your future and your name on the line.

CHAPTER TWENTY-TWO

Staying on Budget and on Time

Keeping your project on time and on budget is instrumental to your success. There are so many elements of land developing that are critical to success that it's hard to pinpoint one that is the most important. Surely, your production schedule and budget rank high on the list of key factors in making a project successful. How will you keep your project on track? Most developers do it with personal attention.

When we refer to personal attention, it doesn't always mean that you will take a regular role in day-to-day activities. Maybe you will have a project manager who will track jobs for you. An accountant might supervise the financial issues as your project develops. Field superintendents might bring you weekly reports to keep you up to speed on the production schedule. Your role might be reviewing reports rather than taking a more hands-on position. Reducing your work by having others tend to daily matters will not reduce your responsibility.

Assuming that you did your advance work properly, staying on budget should be fairly simple. If you have firm contracts that lock in all prices, you shouldn't encounter too many surprises. But there are almost always some surprises to deal with. Maintaining a production schedule depends on so many factors, some of which you just can't control, that it can be difficult to bring a project in on time. Since budgets and schedules are two different aspects of the developing business, let's discuss them one at a time.

Tracking Your Budget

Tracking your budget isn't difficult. Staying within the budget can be troublesome, but it should not be hard to tell where you are from one week to the next. Developers create budgets before they begin site work. The budgets are often on file with the lenders who finance developments. Anyone with experience as a developer expects the budget to come out differently from projections. Occasionally a job comes in under budget. No one gets upset about a job that costs less than expected. But when a job is over budget, there can be a number of problems to deal with. Not only will a developer realize less profit, but he or she may run into trouble with lenders and contractors.

Following the expenditures on a development can be fairly simple. Computer programs can provide you with a ledger that will show all of the expenses to date on a daily basis. Even simple records maintained on ruled paper can keep you up to date on your spending. If you are running a large project, you might want to retain a financial wizard to count your pennies for you. Overall, though, tracking the money being paid is a simple task.

It's not hard to count the money that you've spent, but making sure that you don't spend more money than you have to is somewhat more complicated. Keeping your purse strings tight is wise. If you have cost overruns, you need to be concerned.

Your predevelopment work should include extensive budget forecasting. If you do enough of it and do it right, you should have minimal trouble once fieldwork begins. But don't think for a moment that you will not run into unexpected expenses. You should have entered into your project with some expectation of budget overruns. Most developers factor in some figure to anticipate the unexpected costs. You hope not to need the extra room in the budget, but you hope even more not to exceed it.

Once you are fully engaged in a project, you are committed, even when prices are higher than what you had planned for. This can feel like a helpless situation. If you are way off, it may be extremely serious.

Huge problems are more often than not the result of poor planning. This is why it is so important to pay full attention to all possible expenses during your planning stage. Once your crews are rolling, you are locked into your role

> **PRO POINTER**
> Some overruns are to be expected, but you need to identify why you had to spend more money and evaluate whether the added expense could have been avoided. The lessons learned won't help you on the present project, but they will benefit you on future projects.

and your budget. The best you can do is to limit your losses as best you can, and there may be times when there is little that you can do about your losses. Your best defense is a well-planned offense.

Routine Reports

Routine reports are a good way to keep an eye on your expenses. The reports will tell you where you stand at any given time. However, you should not get too caught up in reports and projections. I knew a developer several years ago who had an entire wall in his office plastered with reports. He would stand there and stare at them for hours. He lost so much time studying his reports that he could probably have started another development. Use the reports as tools, but remember that you have many duties; don't become consumed by the paperwork.

> **PRO POINTER**
> In most cases cost overruns only amount to making less money. If they are so substantial that they reach beyond your profit margin and into your actual expense account, then you have real trouble.

Invoices

When invoices come in, you should check them carefully before paying them. It is quite common for invoices for materials to be more than what you bargained for. Developers who are in a hurry often overlook a few hundred dollars here and there. In the scheme of land developing, $200 is not a lot of money. But, if you overpay bills routinely, the amounts add up to a considerable sum.

Invoices for materials that are higher than what you were quoted should be questioned. Hopefully, you have rock-solid quotes on file to prevent you from having to pay more than you expected to. Sometimes material suppliers simply make mistakes when computing invoices. You have to look for the mistakes. This can be time-consuming, tedious work, but it is very important if you want to stay on budget. Over the years I have caught countless mistakes that would have cost me considerable money. Take the time to review all invoices before paying them.

Invoices submitted by subcontractors should be fast and easy to confirm. Compare the invoices to the contracts and see if the numbers match. Assuming that they do, you are all set. When there is a discrepancy, you should call your contractor to find out why. Don't just pay the bill. Investigate all charges and question any that are not in line with your established contracts.

You should limit your exposure to surprise bills. Don't let contractors get away with tacking on extra costs if you have not agreed to them beforehand. Your contracts should have language that relieves you of any responsibility for cost overruns that are not first negotiated and agreed to in writing.

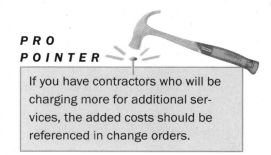

PRO POINTER
If you have contractors who will be charging more for additional services, the added costs should be referenced in change orders.

Price Increases

Price increases sometimes put developers in bad situations. The cost of stone, for example, might go up without notice. You can lock in prices for some period of time, but you may not be able to keep them locked in throughout the full term of a large project. Big projects can take years to complete. When this is the case, the best that you can do is predict price increases as accurately as you can. There will be risk, but there is no other reasonable way to do it.

Contractors should honor their contract prices for the term of a project. They may lose money if they have to offer pay increases to their workers, but the cost should not be passed on to you. If your development can move quickly, you should be able to lock prices in. Your subcontractors will often be responsible for supplying their own materials. When they are, you should have some type of arrangement in your contracts to deal with the risk of escalating material prices.

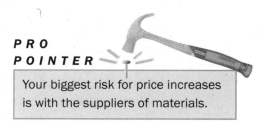

PRO POINTER
Your biggest risk for price increases is with the suppliers of materials.

Staying on Schedule

Staying on schedule is a good way to avoid cost overruns. Projects that run late in their completion schedule normally go over budget. The delay can cost a developer money in various ways. Most projects are financed. Developers budget for a certain amount of finance expense, which is usually called carry cost. When a project runs beyond its intended schedule, the carry costs go up. They can go up quickly and in large amounts. A month or two of additional carry costs on a large project could be quite substantial.

Other reasons that delays cost money can be increases in prices once the locked-in time expires. If your lock-in periods expire before your project is complete, you are susceptible to what could be major price increases. There are plenty of ways for delays to turn into lost money.

Keeping your project on a tight schedule can prove more difficult that you might think. There will be some circumstances that you simply cannot control, such as the weather. A late spring or an early winter could really put you behind. Unusually large amounts of rain could shut you down for weeks. Since you can't control the weather, the best that you can do is plan your job based on past weather records. Once you get into the working stage of site development, you simply have to live with the weather.

Losing three days of production due to rain may not sound like a lot of downtime. But the three days that you lose can throw off work schedules for different contractors. For example, if your clearing crew is set back by three days, it will hold up your excavation crew. If the excavation crew is very busy, your job might get bumped back if it is not ready when it is supposed to be. Your excavation contractor might have to tend to other jobs on their committed start dates. All of a sudden you could be looking at several more days of downtime.

Every time one element of a job is delayed, it can impact other areas. You may have to scurry around letting contractors know that their anticipated start dates have to be moved to later dates. The amount of time spent on rearranging the schedules can take away from your other duties. A ripple effect often occurs when some part of a job is compromised. It is your job to keep your project running as smoothly as possible and to correct problems when they occur.

Your best field supervision is not going to stop the ground from becoming muddy after hard rains. You can think of

> **PRO POINTER**
> Projects that come in late can lose their presale buyers. Lining up buyers for a project is good, but they may be counting on delivery by a certain date. If the date comes and goes without your being able to deliver, the buyers may lose their financing or face higher financing rates.

> **PRO POINTER**
> Supervision is a cornerstone to your production control. If your supervision is set up properly, it will protect you from a number of setbacks that would be likely without supervision.

production control as a part of your supervisory duties, but don't assume that good supervision will automatically result in a timely completion.

Just as supervision is a factor in bringing a development in on time, so is organization. Communication with your subcontractors and vendors is another part of the puzzle. Legwork can speed up a project. By legwork I'm referring to the act of taking an active role in delivering documents and such. The amount of control that you have over your workers can be a factor in getting a job done on time.

Money is almost always a motivator, and it can help to bring projects in ahead of schedule. Backup plans can help you to overcome problems that are slowing your project down. The lure of another project can be enough to keep your contractors grinding out extra hours to complete a present project. Provisions to charge a daily cash penalty to any contractor who is slowing down your project might make a difference. To expand on these issues, let's look at them individually.

Supervision

We've talked enough about supervision to avoid a lengthy section here. When you need to get a project done on time, having someone with authority on your job site is a good way to meet your goal. The person might well be you, your project manager, or your site superintendent. As long as the person has the teeth to back up the barks of command, your project may move along faster. However, if the person is a jerk, you may have any number of problems with your subcontractors and suffer more setbacks than you would have without the supervisor. It takes talent to run a development properly. You have to be strong, firm, and willing to back up your threats. Yet you should strive to gain the respect of your workers so that threats are not needed. Putting the wrong person in a position of power can be a major mistake. And remember that you may be the wrong person. If you are not good when it comes to managing crews, hire someone who is.

The constant presence of an authority figure on a job site can greatly reduce the goofing off that goes on at some projects. However, the patrolling authority may make workers nervous and slow them down or bring about mistakes that might not have happened without the supervision. There can be a thin line between too little and too much supervision. Once you or your supervisor perfects the skills for handling crews in various situations, the hourly presence should make a difference. Even if you are sitting in a climate-controlled site trailer and looking out the door periodically, the knowledge that a boss is on the job should keep things moving along a little better.

Organization

You don't need a massive office and several assistants to maintain good organization. If you are comfortable with computers, a good notebook computer may be your primary means of organization. You can take it with you when you visit a site, you can hook it up to your cell phone to check your e-mail, and you can made notes as needed. A computer can even beep and remind you of your next appointment. For those who are not computer-literate, a spiral notebook can be efficient.

> **PRO POINTER**
> Good organization can help nearly every job come to a close more quickly.

If you have to travel from your office to your site, you may need some travel files. Developers who put site trailers on their projects can keep files in them. However, if you do most of your work out of your truck, you should have copies of files with you. No, you won't need a full filing cabinet in the back of your vehicle. An expandable file folder or a cardboard file box will do nicely. The key is to have copies of active files with you in the event that a question or conflict comes up. For example, if your clearing crew is working your site, have a file with all the documentation for the clearing crew in it. Make sure you have another set of files in your office. Field files sometimes get lost.

Pulling onto a job site and being faced with questions that you cannot answer is bad. It makes you look unprofessional, and it can slow down your job. If there is an assistant in your office whom you can call, that's helpful. But it is better to have active files with you so that you can solve problems on the spot.

Over the years I have found that expandable file folders work very nicely to hold truck files. The folders tuck away behind the seat of a pickup truck with no problem. Cardboard boxes are okay in vehicles like my Jeep, where they can be kept dry and in the back storage area. If your job is small, you may be able to simply put files in your briefcase before leaving your office. Each developer and each project have different needs. The bottom line is to have the

> **PRO POINTER**
> Communication is probably one of the largest factors in successful business operations. Developing land is no exception to the rule. You need to be able to reach your subcontractors quickly and easily. They should be able to contact you in the same manner.

information with you when you need it to keep your project from slowing down or shutting down.

Communication

Cell phones have made communication a lot easier for people who are often on the road and out of their offices. When I contract with a subcontractor, I make sure of what means of communication are available. You should, too. If you deal with an answering machine when you call, you should probably be looking for different subcontractors. Answering machines have their place, but you need contractors whom you can reach quickly when the need arises.

There is much more to communication than being able to contact someone. Once you've made contact, you need a viable means of communicating your needs and feelings. This type of communication is easier with some contractors than it is with others. It's common to have minor personality glitches with some people. While some of your contractors may not be the type of people that you would like to invite over for a party, you have to set aside personal differences if you are going to do business together.

Before I contract with a subcontractor, I conduct a few interviews. These meetings are designed to let us get to know each other. One meeting is not always enough. You and your subcontractors have to decide if you can work together well. Sometimes one of you will feel that you are not compatible in a working relationship. If this is going to happen, find it out early to avoid production problems. One of the last things that you need is subcontractors who don't want to work for you. The meetings take some of your time, but it is far better to spend the time before work begins on a project than it is to scramble to find a replacement subcontractor when your project is in full production.

Legwork

We live in a time where e-mail makes almost instantaneous delivery of documents possible. Fax machines are vital to many businesses. Private delivery services can work wonders in getting packages from one place to another in record time. Even with all of these delivery systems there are times when you simply can't beat old-fashioned legwork. Sooner or later you will run into a situation where a soil sample has to be delivered by a certain time or a lien waiver has to get to a lender in a hurry. If you are available to make deliveries personally, you can save the day. I can't count the number of times that I've acted as a courier for time-sensitive deliveries. Missing a deadline on some deliveries can set a project back by days. This, of course, is not acceptable.

Many cities have courier services, but developers who work out of the city limits can't count on such services. You might be in a position where you will need to meet an engineer on your site and then drive the engineer's findings to some place 50 miles away. Getting the same-day turnaround could make a real difference in your production schedule. It's desirable to avoid legwork whenever you can, but don't be lazy and allow it to cost you valuable production time.

Control

Control is an issue that some people don't like to talk about. People who wear rose-colored glasses like to believe that everyone will do the job in the correct manner, without the need for controlling supervisors. I've been in the trenches long enough to know that this simply isn't true. A lot can be accomplished with a smile and respect, but there are certainly times when control is the only way to win. Most of your control as a developer will come from your contractual agreements.

Subcontractors are entitled to run their own shows, so long as they meet your needs, as per your contract. It wouldn't be right for you to insist on having all red dump trucks on your job site, but you should have the right to demand a certain amount of production. If a subcontractor wants to take two hours off for lunch, you may not be able to do anything but fume about it. But if the subcontractor's action puts your job behind schedule and is not allowable by your contract, then you have some power to speak up firmly.

Control is a tool. It is one you hope will not be needed but will be thankful for when it is. Make sure that your contracts give you plenty of authority. The authority can come in many forms. For instance, you should have a clause in your contract that allows you to replace a subcontractor who is not fulfilling the terms of a contract. A clause that allows you to deduct money from a contract amount to cover losses caused by a contractor is a good idea. Talk with your lawyer and come up with contract elements that will protect you in as many circumstances as you can envision.

Money

Money is almost always a good motivator. With the right bonus structure, you might have some of the happiest subcontractors around and some of the fastest projects in town. I often use bonus programs to stimulate production. It's common for me to take quotes from contractors and add a percentage to them for the payment of possible bonuses. This gives me an existing budget to use, and my contractors love to finish early and receive bonuses.

Backup Plans

Backup or contingency plans should be in place to protect your project from slowdowns. For example, you should have a pool of contractors to call upon if something happens to your first choice. I usually maintain a roster of at least three contractors for each phase of work to be done. If my primary contractor fails to perform, I have two other candidates in position for replacement. This type of advance planning can make a huge difference in keeping a project on schedule.

> **PRO POINTER**
>
> You can use bonus money to get a sluggish project moving even if you don't have a bonus budget. It will be painful to give up your profits, but it will work out to be better than having a late project. In fact, if you consider all the money lost from carry costs and other elements of the probable delay, the payment of bonuses might be saving you money when compared to what you stand to lose.

It's impossible to plan a contingency for every event that might slow a project down. But the more plans you have, the better off you will be. For example, if you have trouble getting a delivery of sewer pipe, what else can you have your backhoe or excavator work on? Think about this type of situation in advance and have an answer ready when you need it.

Another Project

If you have another project coming up as you are nearing completion of an existing project, you might see faster work from your contractors. One reason for this is that they will want to please you so that you will use them on the next project. Another reason is that they will not be worried about what they will be doing when they finish your existing project. This may seem silly, since they are independent contractors being paid by the job, not by the hour. Believe it or not, I have had subcontractors admit to me that they were working slowly because they didn't have another job to go to. Their slow pace didn't earn them any additional money, but it did cost me some in delay factors. Why someone being paid by the job rather than by the hour would work slowly is beyond me. If I didn't have another job to go to, I'd finish up as soon as I could and start prospecting for more work. If you have another project waiting, it could be worth your while to let your contractors know about it.

CHAPTER TWENTY-THREE

Ideas for Environmentally Friendly Developments

What are your ideas for an environmentally friendly development? If you are like a lot of old-school developers, you may have difficulty envisioning a sustainable development. I began dabbling in green development back in the 1980s. A lot has changed since then. Much more is known about suitable principles and practices for creating environmentally friendly developments. Technology has also come a long way over the last couple of decades. Factor in the trend for developers and builders to go green, and you have a bustling opportunity to explore.

If you are new to land developing, you may have some advantage over people like me. You don't have the years of experience as a conventional developer to overcome in your quest to become green successfully. At the pace options change and increase in sustainable living, being young enough to keep up with it could be your advantage over seasoned veterans.

Do you think it costs more per lot to create a desirable sustainable neighborhood development? You might be surprised at what current research and statistics are showing. I was reviewing a case study recently. Factors were compared between conventional developing and sustainable developing. The factors included the following categories:

- Lots created
- Length of streets
- Length of connector streets

- Length of drainage piping
- Components of drainage sections

The results of this study surprised me. Building lots in the green community came in at a cost of about $2,400 per lot less than the lots in the conventional development, based on projections. When the project was completed, the actual saving on each lot was nearly $5,000. And the big surprise to me was that the developer managed to get nine more lots in the green development than would be created in a traditional development. So the developer got more lots at less cost. This particular development started with about 125 acres and managed to reserve over 20 of those acres for green space. These numbers are staggering to me, but they are real.

PRO POINTER If you are new to land developing, you may have some advantage over people like me. You don't have the years of experience as a conventional developer to overcome in your quest to become green successfully. At the pace options change and increase in sustainable living, being young enough to keep up with it could be your advantage over seasoned veterans.

Key Considerations

When you are looking for ideas to create a sustainable community, there are several key considerations. Some of them are noted below:

- Site selection
- Storm-water requirements
- Sanitary drainage systems
- Potable-water systems
- Fire-protection requirements
- Irrigation systems
- General community design
- Suitable density allotments
- Zoning requirements and allowances
- Covenants and restrictions

- Walkways
- Street design and development
- Parking requirements
- Grading
- Landscaping

There are, of course, other considerations. The list I have given you is a good place to start and will get you thinking along the right lines. You should consult local engineers and land planners before you get too far in your development plans.

Partners

Bringing partners into a development deal can allow you to do more with less of your own money or credit. Before letting partners in your camp, be very sure that it is what you really want to do. I have used partners on two occasions and suffered badly from each experience.

If you want to bring good partners to the table, you are going to have to build a strong prospectus for them to review. Identify your goals and objectives. Document what will be needed to realize your dream. Investors and partners will want as many details to slice and dice as they can get their hands on.

> **PRO POINTER**
> If you want to bring good partners to the table, you are going to have to build a strong prospectus for them to review. Identify your goals and objectives. Document what will be needed to realize your dream. Investors and partners will want as many details to slice and dice as they can get their hands on.

Your project is going to need strong credibility to attract the right partners and investors. If you don't already have a successful track record as a developer, you will struggle to find suitable partners. More likely than not, you will become bait for the sharks. Be careful.

Are you willing to relinquish control of your project to see it come into reality? When you bring in partners, it is common to lose some element of control. This is not always the case, but it usually is. Not everyone is willing to march to someone else's tune. If you became a developer to be your own boss, think twice before bringing in partners.

Pedestrians

Pedestrians must be considered when you are planning a green development. Many residents will prefer to walk or ride a bicycle when practical to get to their destinations. To accommodate this desire, you must plan on building suitable systems into your development to meet the needs of future residents. Keep in mind that green builders will look over your plans closely before they commit to buying quantities of building lots. You can bet that experienced green builders will be interested in your routing of pedestrians.

> **PRO POINTER**
> Pedestrians must be considered when you are planning a green development. Many residents will prefer to walk or ride a bicycle when practical to get to their destinations. To accommodate this desire, you must plan on building suitable systems into your development to meet the needs of future residents.

As you plan your pedestrian needs, here are a few factors to keep in mind:

- Identify pedestrian pathways.
- Create paths for bicycles that will not interfere with pedestrians.
- Keep pet exercise areas in mind.
- Design your streets to make them pedestrian-friendly.

General Design

General design factors for a sustainable development will often include the presence of mixed-use buildings. The goal is to reduce travel requirements for homeowners to meet their shopping and other needs. Obviously, your basic design will include green space that results from preserving trees and other natural resources. If clustering homes is a viable option in your region, it will save a lot of land.

Street Solutions

Street solutions for green developments sometimes confuse developers who are new to green options. Traditional streets shed water and create more demand on a stormwater system. Some designers suggest building shorter streets and streets that are narrower than traditional roadways. This works on paper, but I am not convinced that narrow streets are a good solution. It does reduce surface area, but it puts vehicles in closer contact with each other. There can be a case made on either side of the issue, but it is an option to consider.

Grass

Grass is your friend when you go green. Use grassy ditches to channel storm water to suitable locations. Swales covered in grass and natural ground cover are also helpful in the control of storm water. Retention ponds are often the destination for clean storm water that can later be used for irrigation systems.

Ride-Sharing Parking Lots

Incorporate ride-sharing parking lots in your development to encourage car pooling. Use a porous covering for the surface of the parking area to reduce runoff water. Place these lots at strategic points to make them convenient.

> **PRO POINTER**
> Incorporate ride-sharing parking lots in your development to encourage car pooling. Use a porous covering for the surface of the parking area to reduce runoff water. Place these lots at strategic points to make them convenient.

Security is frequently a concern for people leaving their cars in parking lots. Plan on ways to minimize threats to people and property in these communal areas. Lighting is one key. A fenced lot with card-key access is a possibility. A parking attendant is another idea, but you must consider the ongoing expense for this type of service.

Mixed-Use Ideas

I have talked about including mixed-use facilities in your development. This is a broad term. To define examples of mixed-use construction, review the list below:

- Community meeting center
- Exercise areas
- Playgrounds
- Swimming pools

- Tennis courts
- Racquetball courts
- Basketball courts
- Coin-operated laundry facilities
- Office space
- Community library
- Recycling center
- Convenience stores
- Environmentally friendly car washes
- Gas stations
- Daycare facilities
- Grocery store
- Hardware store
- Pet store
- Landscaping store and nursery

There are countless other options to include in the selection of mixed-use facilities. A demographic study of your expected buyers will be of great help to you in defining what the minimum needs for mixed-use space will be.

Preservation

The preservation of existing natural conditions is of key importance. Why pay for expensive landscaping when you can benefit from natural plants already established on your site? Take a hard look at what you have to work with before you make any final decisions. Consult with an environmental engineer to generate a list of options for your particular plot of land. In general, if you have it and can keep it, do it.

> **PRO POINTER**
>
> The preservation of existing natural conditions is of key importance. Why pay for expensive landscaping when you can benefit from natural plants already established on your site? Take a hard look at what you have to work with before you make any final decisions. Consult with an environmental engineer to generate a list of options for your particular plot of land. In general, if you have it and can keep it, do it.

Natural Wind Breaks

Build or keep natural wind breaks to reduce heating and cooling needs in the homes that you are developing lots for. Deciduous trees are those that lose their leaves before winter. This allows sunlight to reach homes during the winter. If you expect to have solar-equipped homes in your development, and you should, this is an important consideration. Leaves on the trees in summer will shade homes and reduce cooling expenses.

Evergreen trees block more wind in winter, but they also block sunlight. A mixture of tree species is the right recipe. Take the lot orientation into consideration when deciding on what types of wind breaks to create. Low-growing trees have advantages in allowing light into a home. Tall trees block more wind. Mix and match as needed. A landscape architect can be of substantial help in the decision of what to plant and where to plant it.

> **PRO POINTER**
>
> Build or keep natural wind breaks to reduce heating and cooling needs in the homes that you are developing lots for. Deciduous trees are those that lose their leaves before winter. This allows sunlight to reach homes during the winter. If you expect to have solar-equipped homes in your development, and you should, this is an important consideration. Leaves on the trees in summer will shade homes and reduce cooling expenses.

On-Site Wastewater Treatment

On-site-wastewater-treatment systems are something that you might not think of as a conventional developer. If municipal sewer services are available, you may be required to tap into them. Check this situation with your experts and your local code-enforcement office. If you have the option to use a private wastewater system, it could be beneficial to your development. There are, however, drawbacks to this type of system, so let's explore the topic.

Before a private wastewater system can be considered seriously, you must determine if there is a suitable site on your property to house the system. Soil engineers are the experts to turn to for this determination.

Buyers in sustainable communities may like the option of having wastewater treatment occur naturally in the earth. The flip side to this is that some residents will be repulsed by the use of a sewage system on site. Another disadvantage of a private system is routine maintenance and management. These costs have to be paid by someone.

When you run the financials to compare a private system to a public sewer system, you have to weigh all of the associated costs. You can count on routine maintenance and management with a private system. What will you do if a drain field clogs and has to be replaced? Will insurance cover the expense? It depends on your coverage. What will you do if odor becomes a problem in your community? As a master plumber, builder, and developer, I would tap into a municipal sewage system if I had the opportunity. There are people who see the benefit to a private system, as do I, but the risk outweighs the reward in my personal opinion.

Covenants and Restrictions

As a developer, you will be responsible for setting the covenants and restrictions for your development. These elements maintain a certain quality and value to your development. You will have to decide how far to take your rules and restrictions. Keep in mind that more rules could result in fewer sales. However, good rules can fetch higher prices for lots. It is a bit of a gamble. If I were rolling the dice, I would go for a quality development and count on it to grow in value as it is built out.

> **PRO POINTER**
>
> As a developer, you will be responsible for setting the covenants and restrictions for your development. These elements maintain a certain quality and value to your development. You will have to decide how far to take your rules and restrictions. Keep in mind that more rules could result in fewer sales. However, good rules can fetch higher prices for lots.

Your attorney can help you to establish the legal requirements for builders and homeowners who will be active in your community. It will be your job to set forth a list of ideas for consideration. A lawyer can get the job done, but the developer has to lay out the game rules. Here are a few examples of what you might place in your covenants and restrictions:

- All homes built must comply with approved "green" standards.
- Building materials must be durable and require minimal maintenance.
- Green framing techniques must be used in the construction of buildings.
- Gutters and downspouts must be run to approved drainage systems.
- Irrigation systems are required to use recycled water.

Ideas for Environmentally Friendly Developments

The list of potential restrictions could go on and on, but this should give you enough of an idea to start you thinking about what you would want from your sustainable community.

Collect Brochures

Collect brochures from other green developers. Their sales agents will be happy to provide you with a host of materials pitching their properties. Study these brochures closely and borrow ideas from them. Before you do this, come up with your own plans so that you will not limit your personal creativity. Once you are at a stalemate in your development plan, visit the brochures for additional fresh ideas. If you borrow the ideas of 10 developers to incorporate with your own, you are likely to have a winning project.

Talk with Bankers

Talk with bankers. You are probably going to need them to finance your project. In this case, you are not seeking their approval for a loan. Instead, you are asking what buyers want in a green development. Bankers will often know what people like. It is common for homebuyers to brag about what they are getting for their money when they are signing off on their mortgages.

Real-Estate Professionals

Real-estate professionals are always willing to talk with active land developers. Take advantage of this. Ask them what they believe makes a sustainable project successful. Make notes, but take what they say with a grain of salt. Ask for data on comparable sales to substantiate their claims. This will separate what real-estate experts call puffing from facts. Puffing is not a lie, but it is a stretch of the truth. You want hard facts and numbers to work with.

> **PRO POINTER**
>
> Real-estate professionals are always willing to talk with active land developers. Take advantage of this. Ask them what they believe makes a sustainable project successful. Make notes, but take what they say with a grain of salt. Ask for data on comparable sales to substantiate their claims. This will separate what real-estate experts call puffing from facts. Puffing is not a lie, but it is a stretch of the truth. You want hard facts and numbers to work with.

Interview Development Professionals

Interview development professionals. I am talking about landscape architects, engineers, surveyors, and such. You will need these services when you launch your project, so you might as well get more mileage out of your relationship. Turn the interview for using their services into a brain-picking expedition for ideas. Most of them will be happy to tell you war stories and brag about their successes. What has worked for them in the past could work for you in the future. This is free advice that could prove to be invaluable.

> **PRO POINTER**
>
> Interview development professionals. I am talking about landscape architects, engineers, surveyors, and such. You will need these services when you launch your project, so you might as well get more mileage out of your relationship. Turn the interview for using their services into a brain-picking expedition for ideas. Most of them will be happy to tell you war stories and brag about their successes. What has worked for them in the past could would for you in the future. This is free advice that could prove to be invaluable.

Reach Out

Reach out to potential customers. Set up a meeting event or seminar for green builders. They are likely to be your customers. Invite them to a barbeque, a town-hall dinner, or something along these lines. Survey them on what they want from you as a developer. You will be meeting potential buyers, having a chance to impress them, and gaining very valuable data from what they expect out of your development. It will cost you a few bucks to set up the meet and greet, but it should be well worth it.

Bottom Line

The bottom line is to get creative. Talk to people in appropriate places to see what they would want in a green development. What is an appropriate place? It could be almost anywhere, but here a few examples:

- Greenhouses and nurseries
- Building-supply stores

- Green-building product stores
- Environmental-club meetings
- The section on green building in your local bookstore

If you put your mind to it and want them badly enough, you can find the answers you need to define a buyer-friendly subdivision. Research is frequently the key to your ultimate success. There are plenty of pitfalls, but the more you know before you start your journey, the more likely you are to be successful.

CHAPTER TWENTY-FOUR

Going from Green Developing to Green Building

What are the benefits of going from green developing to green building? Money and control are two large benefits of expanding your services. When you are your own developer and your own builder, you have complete creative control over your development. Any developer can use covenants and restrictions to cast the die for a development. But being your own builder takes the control a step farther. This type of control gives you the maximum power in creating a successful, sustainable community.

The financial rewards of being a developer and builder can be very substantial. To expand on this, I will give you a sample from my past experience as my own developer and builder. Consider this case study carefully.

20-Acre Mini-Estates

Some years back I was involved in a 200-acre development. Technically the acreage was a bit more than 200 acres, but once the road was installed we were left with 10 lots. To minimize cost and to keep the development desirable to friends of nature, we cut the 200 acres into 10 parcels that contained 20 acres each. The road was very simple; it ran in a straight line through the middle of the acreage.

This development was very simple to create. Expenses for the development were minimal. Once the land was cut into lots, I built a model home on the first parcel. We were able to sell all the lots from a single model home. Profit from the land development doubled the cost of the land. This in itself was great, but the deal got better.

The houses that we were building yielded about a 20-percent income from the total price of the house and lot value. To keep the math simple, say that the house and lot appraised for $200,000. The builder income was around $40,000 per house—in addition to doubling our money on the land as a developer.

Now consider that we used an in-house sales team to do the selling. Real-estate brokerage fees for new construction at the time were 5 percent of the selling price. Using our sample numbers, this amounted to a $10,000 commission per house. With ten houses, this amounted to $100,000. Not a bad bonus for a development that sold out in one summer. Now let's do all the math using the sample numbers:

- Ten lots were developed with a value of $40,000 each. They cost $20,000 each. This left a $200,000 income for less than one year of effort as a part-time venture.
- Ten houses gave an income of $400,000, and we could produce each house in less than 90 days, all of them in one building season.
- Ten houses provided $100,000 in money saved on real-estate commission. Granted, there were advertising expenses and a sales representative to pay, but that probably cost less than $50,000.
- The total estimated income for less than one year of effort was $650,000.

As you can see, the income potential is very good when you combine developing and building. You can triple your income by becoming a builder for your own developments.

Before you throw down this book and run out to buy a hammer, let me tell you a little more about my experiences in this world. Projects can look great on paper and flop in reality. Land developing and speculative building are risky. The old saying about the greater the risk, the greater the reward does tend to hold true. This type of business is not a pot of gold at the end of a rainbow nor a guarantee of getting rich quick. To succeed, you have to evaluate every opportunity on its own merits.

What Does It Take to Become a Builder?

What does it take to become a builder? It depends on where you will be working. Some states require that you pass an exam on building codes and practices. A few places require builders to have field experience as foremen before they can be licensed. Other states only require a builder to obtain a simple business license. A bond might be required in some states.

To determine the requirements in your region, check with you local licensing authorities. This can often be done on the Internet. It is common for jurisdictions to post the requirements for licensing in their area on the Web.

Do I Have to Be a Carpenter?

Do I have to be a carpenter to become a building contractor? Generally speaking, the answer is no, you don't. As I mentioned earlier, some licensing agencies do require field experience. Even when this is the case, there are ways to get around the requirements. For the moment let's assume that you are not required to have field experience to become licensed as a builder. I will teach you the end-around move shortly.

> **PRO POINTER**
>
> What does it take to become a builder? It depends on where you will be working. Some states require that you pass an exam on building codes and practices. A few places require builders to have field experience as foremen before they can be licensed. Other states only require a builder to obtain a simple business license. A bond might be required in some states.

Being a building contractor is a business. You don't have to install roofing, siding, plumbing, or any other form of manual labor in the construction process to be a successful building contractor. There are key qualities that will enhance your ability as a contractor. Here are a few of them:

- Be very well organized.
- Understand the building process from the ground up. You can do this by reading or taking classes.
- Understand the process of working with subcontractors. Again, you can gain a lot of knowledge on this topic from reading professional reference books.
- Ideally, get a handle on marketing and sales strategies.
- Establish good relations with at least two financial institutions that offer construction loans and permanent mortgages.
- Keep your overhead expenses as low as you can while maintaining a professional appearance.
- Learn how to do your own estimates on labor and material costs. This can be done with the help of estimating guides.
- Build a strong, dependable team of subcontractors to work with.

Are you still wondering how to bend the rules in jurisdictions where field experience is required to be licensed? It is really rather simple. Hire a field superintendent or lead carpenter who has years of experience to work for you. In this situation, you are putting up the money and the employee is putting up the field experience to carry the license. This is not an uncommon form of so-called partnership.

What Does a Builder Do?

What does a builder do? I have worked in construction for over 30 years. During these years I have owned and run plumbing, remodeling, real-estate, building, and land-development businesses. Many times they all existed at the same time. I used to build about 60 single-family homes a year. Procedures for builders vary, but I can give you insight from my personal experience and my knowledge of fellow builders over the years.

> **PRO POINTER**
> Building a model home is not essential, but it surely helps. Try to get your lender to agree to at least one model home. Having three models to sell from is better when you have a larger development. Keep in mind that you will have carry costs on the interest required for the construction loan for the model homes. This has to be factored into your earnings and cash-flow needs.

When you are your own developer, you don't have to worry about finding suitable lots, negotiating take-down schedules, and other elements of acquiring positioning in the subdivision. You can generally roll your financing into one deal that covers the land development and the constructions loans that are typically used to build both speculative and custom homes.

Building a model home is not essential, but it surely helps. Try to get your lender to agree to at least one model home. Having three models to sell from is better when you have a larger development. Keep in mind that you will have carry costs on the interest required for the construction loan for the model homes. This has to be factored into your earnings and cash-flow needs.

Subcontractors will be your biggest hurdle in many ways. Finding licensed, insured, dependable subcontractors at competitive rates can be a daunting job. If you are going to be a building contractor, you will need professionals from surveyors to landscapers and every other skill in between. Don't take this lightly. It is a necessary and often difficult task to master.

It is logical to assume that you can call plumbers and ask them to bid your work quickly and easily. Don't count on this. Dealing with subcontractors can be your downfall. Make sure that you have solid subcontractors in place before you launch into a building project. One way to do this is to join associations and organizations for builders, where you can network with other builders and get leads on good subcontractors.

Get your office, even if it is in your home, well organized before you hang out your shingle as a builder. You will need detailed files and records. There are many good computer programs available for builders, and these are a good option. But file folders, legal pads, and ink pens will still get the job done for those who shy away from an electronic office.

Build your sales team early. If you are not going to have an in-house sales team, interview real-estate brokerages long before you need them. They can help you in pricing and marketing. It is worth having your blueprints appraised by bank-approved appraisers before you build the houses. By doing this you know what the market value is and you remove the guesswork that some builders go on.

Building is a big task, but it is really a manner of being a good manager who is well organized and willing to work long hours. If this fits your personality, you may be a viable builder. There is inadequate space here to discuss all that you need to know to be a successful builder, but there are other books available to prepare you for the expansion of your business. In fact, I have recently written a companion book for McGraw-Hill titled *Be a Successful Green Builder*.

Getting into the building side of the business can clearly increase your income potential. It is not without risks or commitments, but for many people it is a worthwhile venture. We are at the end of our journey with this book. I wish you the very best in developing sustainable developments that we can all benefit from.

APPENDIX ONE

Glossary of Green Words and Terms

Absorption: The process by which light energy is converted to another form of energy, such as heat.

ACH (Air changes per hour): The movement of a volume of air in a given period of time. An ACH rate of 1.0 means that the volume of air will be replaced in a period of one hour.

Acid leachate: Water that is made highly acidic after seeping through landfills; it may be harmful to fish habitats and drinking-water supplies.

Active system: Heating, cooling, and ventilation systems that condition the air supply in buildings by using electricity or gas power.

Adaptable buildings: Buildings that may easily be remarketed, reconfigured, or retrofitted in order to meet the changing needs of maintenance crews, occupants, and the surrounding community.

Adsorption: The process in which the molecules of a gas, liquid, or dissolved substance adhere to a surface.

AFUE Annual fuel-utilization efficiency: A measurement of the efficiency of gas appliances, the ratio of energy output to energy input on an annual basis.

AFV (Alternative-fuel vehicle): Any vehicle powered by fuels other than gasoline.

Agricultural by-products: Materials left over from agricultural processes, such as shells and stalks, now being used for building materials.

Air infiltration: Undesired air leakage in a building shell. Air leaking out is referred to as exfiltration.

Air retarder/air barrier: Materials installed around the building frame to prevent and reduce air infiltration, used to increase energy efficiency by keeping out air that may be too cold, hot, or moist.

Albedo: The ratio of light falling on a surface to the amount of light reflected off the surface. Roofing materials with different ratios (high or low) of albedo produce different results.

Appraisal value: Generally substantiated through comparison with other properties, the estimated value of a property.

Allergen: Any substance that a person may show an allergic reaction to.

Attic venting system: Ventilation devices installed to allow fresh air in and allow the exhaust of air out to control air quality in an attic. The use of a continuous soffit vent with a continuous ridge vent is most effective, because it allows for even air flow along the underside of the roof while exhausting the hottest air at the highest point in the attic.

Autoclaved cellular concrete: A molded mix of concrete, sand, lime, water, and aluminum, which is then steam-cured for strength. Benefits of this process include a noncombustible, easily worked product.

Backdrafting: Combustion gases entering the living space instead of being properly drawn up an exhaust pipe; this happens as a result of depressurization, sometimes caused by exhaust fans or furnace heat.

Backflow preventer: In accordance with many building codes, an anti-siphoning device used on water pipes to prevent contaminated water from backing up into the water system.

Balance point: The temperature at which a building's internal heat gains are equal to the building's loss of heat to the surrounding environment.

Ballast: A magnetic or electronic device used to provide the stabilizing current or starting voltage of a circuit.

Bioaccumulants: Found in contaminated air, water and food, substances that are very slowly metabolized or excreted by a living organism, causing buildup over time.

Biodiversity: An ecosystem that contains a great variety of species and a complex web of interactions between them. Human influence, mainly related to resource consumption, greatly reduces biodiversity and heightens risk of catastrophic disruption to these systems.

Bioengineering: Using a combination of living plants and nonliving materials to stabilize slopes and drainage paths.

Biological wastewater management: Wastewater purification programs that are based on natural processes of wetland environments and powered by sunlight and microscopic living organisms.

Biomass: Any natural material—animal manure, wood and bark residues, or plants and plant materials.

Biomass energy: Energy released from biomass as it is used or converted into fuel.

Blackwater: Dirty water from sources such as toilets, kitchen sinks, and washing machines, which may be contaminated with harmful bacteria or microorganisms.

Blown-in batt: A method that uses a high-power blower and a fabric containment screen to install loose insulation into wall cavities.

Borate-treated wood: A mineral derived from borax that is less harmful than most other wood treatments. Treating wood with borate makes it more resistant to moisture and termites.

Brownfields: Areas where environmental contamination inhibits the redevelopment or expansion of old commercial and industrial sites.

BTU (British thermal unit): A measurement of heat energy, about the amount of heat needed to raise the temperature of a pound of water by one degree, or the amount of energy released by lighting a match.

Building codes: Municipal ordinances pertaining to health and safety, used to regulate construction and occupancy rules.

Building ecology: The physical environment of the interior of a building. Air quality, acoustics, and electromagnetic fields are key issues of building ecology.

Building envelope: The enclosed space of a building, closed in by walls, windows, roofs, and floors, through which energy is transferred to and from the interior and exterior.

Built environment: Any human-built structure in contrast to the natural environment.

Caliche: A common roadbed material made of calcium-carbonate-rich soil, which hardens without firing.

Capitalization rate: The rate at which future income flow is converted into a present value figure, expressed in a percentage.

Carbon dioxide (CO_2): Composed of one carbon atom and two oxygen atoms, a molecule formed through the processes of animal respiration or in the decay or combustion of organic matter. It is an atmospheric greenhouse gas, and plants absorb it in the process of photosynthesis.

Carbon monoxide (CO): Formed as a product of the incomplete combustion of carbon, a very toxic gas made of carbon and oxygen. When burned, it shows a blue flame and creates carbon dioxide.

Carrying capacity: The amount of a particular product that may be used without depleting the source or degrading dependent life forms.

Cellulose: The fibrous part of a plant, currently used to make paper and textiles, and may also be used to make building products such as insulation.

Cementitous: Having the properties of cement, the primary bonding agent in concrete.

Certified lumber: Lumber that is harvested from operations certified as sustainable.

CFCs (chlorofluorocarbons): A number of compounds used in refrigeration, cleaning solvents, and aerosol propellants that contain carbon, chlorine, fluorine, and hydrogen. CFCs are also used in the manufacturing of plastic foams and have been linked in recent years to the massive depletion of the ozone layer.

Change order: A permission request by a contractor or architect to make changes to an approved plan.

Cistern: A tank, above or below ground, used to hold fresh water.

Co-product: Anything left over from material processing that may be further processed and converted to usable materials.

Color temperature of light: The color appearance of a light, either cool or warm.

Combustion gases: Gases that are created through the process of burning, such as carbon monoxide. Gas appliances in a home may produce these gases, so proper ventilation is important.

Community: A group of several different species living in a defined habitat, which they share while also maintaining independence from each other.

Compact fluorescent lighting: A fluorescent light that is made to fit in an Edison light socket, used as an efficient alternative to incandescent lighting.

Composting: The process used to control decomposition of organic materials into a humus-like material that may be used as an organic fertilizer.

Composting toilet: A toilet that processes waste material into a material that may be used as a soil amendment.

Conduction: The transfer of heat through solid materials that are in contact with each other.

Conductor: Any substance or object able of transmitting heat, energy, or sound.

Constructed wetlands: Used to treat runoff and wastewater, a variety of systems designed using natural wetlands as a model.

Construction waste management: A term used to describe construction or demolition strategies that encourage recycling and reuse of materials.

Cooling/heating load: The amount of heat/cool air needed to offset a deficit/overage of the other.

Covenants: Worked into a deed, agreements that allow or disallow certain activities and usages on the deeded property.

CRI (color rendering index of light): When objects are illuminated by electric light, they will appear differently depending on the quality of the light. Rated on a scale of 1 to 100, objects will appear closer to their actual color with higher numbers and more distant with lower numbers.

Cross ventilation: The proper sizing and placements of doors, windows, and walls to cool a building by using natural breezes and air flow.

Critical zone: A location within a building that has numerous contaminant sources, thus requiring proper ventilation control to remain comfortable for occupants. Critical zones include: cafeterias, washrooms, auditoriums, smoking rooms, or any room where occupancy changes rapidly.

Cullet: Waste glass that has been crushed and returned for recycling.

Daylighting: The use of natural lighting for interiors through the placement of windows, skylights, and reflected light.

Degree days: A rough measure used to estimate the amount of heating required in a given area; the difference between the mean daily temperature and 65 degrees Fahrenheit.

Demand hot-water system: A hot-water heater designed to supply hot water instantaneously as opposed to storing preheated hot water in a tank. This system allows for the elimination of energy wasted to keep stored water warm, minimizes the amount of water wasted waiting for water to get warm, and helps to reduce conductive losses while traveling through pipes.

Design conditions: The interior and exterior environmental boundaries set for air conditioning and electrical design of a building.

Design temperatures: Temperatures used to base energy calculations on; calculations of extreme highs and lows established for different cities and different seasons.

Direct sunlight: The portion of daylight that arrives directly from the sun, without any diffusion.

Diurnal flux: The difference between daytime and nighttime temperatures, measured in degrees Fahrenheit. A diurnal flux of 25 degrees or more is considered an arid climate and is suitable for mass building construction.

Drip irrigation: A low-pressure, above-ground tube irrigation that constantly releases small amounts of water at the base of a plant.

Drought tolerance: The ability of a landscape plant to survive and function properly during drought conditions.

Earth sheltering (earth berming): Building a shelter below ground level, where soil temperatures are steadier and closer to the temperature desired, leading to 40-60 percent higher efficiency in heating and cooling systems.

Eave: The extension of the roof over the edge of a building, which serves to protect the structure's sides from the elements.

Ecology: From a biological standpoint, the study of relationships among living organisms and their surrounding environment. From a sociological standpoint, it is the interactions between human groups and their surroundings, also considering material resources and the consequential social and cultural patterns created.

Ecosystem: The delicate balances of a system of natural elements on which a habitat depends for survival.

Ecotourism: Tourism that partners with conservation efforts to preserve and protect the natural and cultural attractions of destination areas.

Edible landscaping: Using edible vegetation in a landscape design, such as fruit trees and fruit-bearing shrubs.

Efficiency: The ratio of energy input to useful energy output of a device.

Electricity: A flow of electrical power or a charge of energy created through friction, induction, or chemical change caused by the motion or presence of elementary charged particles.

Electronic ballast: A type of ballast for fluorescent lighting that can reduce flicker and noise, while also increasing efficiency.

Embodied energy: The energy needed to grow, harvest, extract, manufacture, refine, process,, pack, ship, install, and finally dispose of a particular object, product, or building material.

Emissivity: The ability of a material to convey far-infrared radiation through an air space. Materials with a low emissivity rating have a poor ability to do this and can be useful for blocking heat and controlling hot-air distribution in hot climates.

Encapsulation: A protective coating applied to asbestos-containing material in order to prevent the release of harmful fibers into the air.

End-use/least-cost: Focusing on the end user's needs, a decision-making tool to achieve the greatest benefits at the least cost in financial, social, as well as environmental terms.

Energy: The ability to do work; Btus and kilowatt-hours (kWhs) are common English units of energy measure.

Energy conservation: Efficiency of energy use, transmission, production, or distribution that creates a reduction in energy consumption while providing the same or higher levels of service.

Energy or water efficiency: The act of using less energy or water to perform the same basic tasks. A device is energy-efficient when it displays comparable or better quality of service while using a smaller amount of energy than other available technologies.

Engineered wood: Wood products that are reconstituted to result in various strengths and quality, producing a consistent product with less material.

Erosion: The wearing down of land surfaces caused by wind and water, the effects of which may be increased by land-clearing procedures related to farming, industrial or residential land development, logging, or road building.

ERV (energy-recovery ventilator): A type of mechanical equipment that uses an air-to-air heat exchanger in conjunction with ventilation systems to control air temperature and humidity levels in a building.

Evaporative cooling: Limited to arid climates, a cooling strategy that employs water evaporation into a hot, dry air stream in order to cool it.

Expanded polystyrene: A rigid insulation material, frequently made of recycled product and with CFC-free processing, made by heating pentane-saturated polystyrene pellets into various densities designed for different uses.

Feedstocks: Raw materials used to manufacture a product, such as the oil or gas required to produce a plastic.

Fenestration: A term that refers to any arrangement of openings in a building that admit daylight and any architectural elements that may affect light distribution.

Fluorescent lamp: A lamp that produces light by passing an electric arc between two tungsten cathodes in a tube filled with low-pressure mercury vapor and other gases.

Fly ash: Ash residue formed during high-temperature combustion processes, useful in the mixture of concrete.

Formaldehyde: Commonly used as an adhesive in wood products, this pungent-smelling, colorless material can be highly irritating if inhaled and is a probable human carcinogen.

Fossil fuels: Fuels, such as oil, coal, and natural gas, that take millions of years to form from the altered remains of once-living organisms.

Geotextiles: Fabrics engineered to be used in the soil for water and erosion control.

Geothermal/ground source heat pump: A heat pump that uses underground coils to collect and transmit heat to a building.

GIS (geographical information system): Used in the evaluation of probable building locations, landscaping, and land use, a technology that connects databases of information such as soil content, hydrology, and plant and animal habitats with mapping programs.

Glazing: Window glass, clear plastic films, or any type of transparent or translucent covering that protects from weather while still allowing light to pass through.

Global warming: As a result of greenhouse effects, a gradual increase of the earth's temperature over a long period of time.

GRAS (generally regarded as safe): A term used to describe products that have been used for many years without the occurrence of toxic side effects.

Graywater: Water that has been used for clothes washing, sinks, and showering and is suitable for reuse as subsurface irrigation in yards. Water from the kitchen sink and toilet is excluded from this category.

Green development: A development approach that is based on interconnected elements of environmental responsiveness, resource efficiency, as well as sensitivity to the existing community.

Greenhouse gas: A number of heat-trapping or radiatively active trace gases found in the earth's atmosphere that absorb infrared radiation, including water vapor, carbon dioxide, methane, ozone, CFCs, and nitrogen oxides.

Green roof: A green space integrated with or located on the roof of a building, consisting of a layer of living plants in a growing medium with a drainage system. It contributes to sustainable building strategies by reducing storm-water runoff, modulating temperatures in and around the building, and providing habitat for wildlife and functional open space for people.

Green wash: A false claim that a product or organization is environmentally sound, also known as faux green.

Habitat: A specific environment that organisms or biological populations rely on to live and grow.

HCFCs (hydrogen chlorofluorocarbons): A contributor to ozone depletion, but only one-twentieth as potent as CFCs.

Heat-island effect: A rise in ambient temperature in large, paved areas. Placed strategically, trees and landscaping elements can help reduce this effect while also substantially lowering cooling costs.

Heat pump: Used for heating and cooling, a mechanical device that moves heat from one location to another. Heat pumps draw from multiple sources of heat, including both air and water sources.

Heat-recovery ventilator: A device that, by forcing outgoing air past incoming air, can help retain temperature control and cut energy use for heating and cooling systems by 50-70 percent; also called an air-to-air heat exchanger.

HEPA (high-efficiency particulate-air) filter: A high-quality air filter, usually exceeding 98 percent atmospheric efficiency, typically used in clean rooms, surgeries, and in other special applications.

High-heeled truss: A roof-truss design with a space for insulation near the eaves. Other roof truss designs constrict the amount of space for insulation in this area.

High-mass construction: The passive building strategy of using heat-retaining materials, such as adobe and masonry, to help moderate diurnal temperature swings.

Household hazardous waste: Products used and disposed of in a residential setting that may be hazardous, including paints, stains, varnishes, solvents, pesticides, and any materials or chemicals that are corrosive, toxic, flammable, or explosive.

Humidistat: A device used to measure relative humidity.

Human health risk: The probability that a given exposure or series of exposures may have caused damage or could cause damage to an individual's health.

HVAC: Heating, ventilation, and air-conditioning system.

Hydronic heating: A system that heats space by using water circulated in a radiant floor or baseboard system or a fan-coil or convection system.

Impervious cover: A watertight barrier that covers the ground and does not allow water to seep into the soil below. Used correctly, it can prevent nonpoint source pollution at construction sites.

IAQ (indoor air quality): The health effects and cleanliness of the air in a building, highly influenced by the release of compounds into the space by various materials, microbial contaminants, and carbon-dioxide levels. Choice of building materials, cleaning procedures, and ventilation rates all highly determine IAQ.

Infill: A form of development that helps prevent urban sprawl and promotes economic revitalization by developing on empty lots of land already contained in an urban area, as opposed to on undeveloped lands outside of the established urban zone.

Infrastructure: Roads, highways, sewage, water, emergency services, parks and recreation, or any other service of facilities provided by a municipality or private source.

Insulation: Typically installed around living spaces to regulate and improve heating controls, a variety of materials, each with an R-value to define the resistance it has to heat flow. Materials with a higher R-value are more insulating and can be used effectively to slow the flow of heat.

Integrated design: Interactions between design, construction, and operations taken into consideration in an attempt to heal and protect damaged environments while reintroducing production of healthy food, clean water, and air into biologically healthy human communities.

Kilowatt (kW): A measure of electrical power, one kilowatt is equal to 1,000 watts.

Kilowatt-hour (kWh): A measurement of energy that uses the amount of power multiplied by the time of use to equal one unit.

Land stewardship: Managing land and resources in such a way as to promote sustainable and restorative actions.

Latent heat: The amount of heat required for a material to change phases (liquid to gas) without altering its temperature.

Latent load: Resulting from thermal energy, the cooling load released when air moisture changes from vapor to a liquid state. In humid, hot climates, cooling systems must have proper capacity to handle the latent load while still maintaining a comfortable climate for inhabitants.

Lead ventilation: The ventilation of a new, unoccupied building space to dilute contaminates from construction and HVAC systems to acceptable levels before occupants arrive.

Lease: A contract that allows a tenant to possess a domicile for a specified period of time while paying rent to a landlord.

Leichtlehm: Typically used in making walls, a mixture of straw and clay, moistened and molded between forms, that hardens into a strong material.

Life cycle: The sequential processes and interlinked stages of a product, beginning with its extraction and manufacturing and ending with recycling and waste-management operations.

Life-cycle assessment: The process of evaluating the cost of a product that considers all steps of a material's life cycle, including extraction and processing of raw materials, manufacturing, transportation, distribution, use, maintenance, reuse, recycling, and disposal.

Light: The visual perception of radiant energy.

Light construction: The construction of a building using materials of lower densities, reducing the capacity to store heat.

Light shelf: A daylighting strategy that bounces natural light off of a shelf below a window and onto a ceiling, bringing light deeper into the inside space.

Light-to-solar-gain ratio: The ratio of solar-heat gain to the ability of glazing to supply light.

Lignin: In wood, the naturally occurring polymer that binds the cellulose fibers together.

Linoleum: A natural and durable flooring material, primarily made from cork, that may also be used for other applications such as countertops.

Locally sourced materials: Obtaining materials from a defined radius to help lower impact by reducing transportation and energy usage while also supporting local economies.

Louvers: A set of baffles used to absorb unwanted light, shield light sources from view at certain angles, and allow for selective ventilation.

Low-emissivity windows: Glazing with special coatings used to allow most of the sun's light radiation through while preventing heat radiation from doing so.

Lumens: The amount of light given off by a light source.

Mass transit: The movement of people or goods from starting point to destination by use of public transportation systems such as bus, light rail, or subway.

MDF (medium-density fiberboard): A composite wood fiberboard, typically used in cabinetry and other interior applications, that may sometimes contain urea formaldehyde, contributing to poor indoor air quality.

Methane (CH_4): A colorless, odorless gas, nearly insoluble in water, that burns a pale, faintly luminous flame and produces water and carbon dioxide (or carbon monoxide if oxygen is not present).

Microclimate: The specific climatic conditions of a building site, affected by the site's geography, topography, vegetation, and proximity to bodies of water.

Mineral fibers: Used in insulation, glassy materials that have been melted and spun to create very fine fibers that are an inhalation hazard.

Mixed air: In an HVAC system, the mixture of outdoor air with return air, conditioned and filtered so it can be used for supply air.

Mixed-use development: A development strategy that integrates multiple revenue-raising uses into a single or multi-building plan, generally including housing, retail, and office space.

Mortgage: A written contract that requires real estate as collateral for the payment of a specified debt.

MSDs (material-safety data sheets): Documents required by OSHA to be provided by the manufacturer of products that are potentially hazardous. They contain information about potentially hazardous airborne contaminates, warnings, inspection tips, health effects, odor description, volatility, combustion contaminants, reactivity, and cleanup and disposal procedures.

Mulch: A layer of organic material (straw, wood chips, leaves, etc.) spread around plants to retain moisture, prevent growth of weeds, and enrich the soil.

National Fenestration Rating Council: A council that rates window models in a variety of areas such as light transmittance and energy efficiency.

Native vegetation: A plant that lives in a specific region naturally, not due to human cultivation or intervention.

Neotraditional planning: A type of planning, based on nineteenth-century American town plats, that aims to minimize automobile use and strengthen a sense of community by centering the plan on a town center and public open spaces.

New urbanism: An urban-planning movement that emphasizes inner-city revitalization and the reform of the American suburb community. Points emphasized include diversity of use and population; a focus on pedestrian-friendly access, and a well-defined public realm.

Nighttime ventilation: An energy-conserving building strategy that utilizes cool nighttime air to flush the building and minimize the next day's cooling-energy load.

Nonpoint source pollution: Types of pollution that are difficult to link to one target source, typically pollutants of water.

Nonrenewable fuels: Fuels that are not easily manufactured or 'renewed,' Nonrenewable fuels, such as oil, natural gas, and coal, could one day be completely exhausted.

Nonrenewable resources: Resources that are limited and are used up much more quickly than they can be produced. These resources face exhaustion.

Occupancy sensor: A sensing device used to control lighting, ventilation, and heating settings based on whether the space is occupied or not.

Off-gas/out-gas: Emission of fumes into the air from new products such as new paint, carpeting, and various building materials. Many of these chemical compounds may be unpleasant and harmful to your health.

On-demand hot water: See demand hot water system.

Operating costs: All costs related to management of a property, including: maintenance, repairs, and operation of the property, utilities, insurance, and property taxes.

Organic matter: All natural materials of plant or animal origin.

Orientation: The alignment of a building to the directions of the compass and the sun.

OSHA (Occupational Safety and Health Administration): A federal agency created in 1971 for the purpose of preventing work-related injuries, illnesses, and deaths.

Outdoor air supply: Air that is brought into a building from outside.

Ozone (O_3): A molecule configured from three oxygen molecules. Ozone is poisonous at the earth's surface, but in the stratosphere the ozone layer protects the earth from harmful ultraviolet space radiation.

Particulate pollution: Pollution composed of tiny liquid or solid particles suspended in a water supply or the atmosphere.

Passive building design: Configurations of buildings that utilize natural and renewable resources, such as sunlight and cool air.

Passive cooling: A cooling strategy that combines the use of shaded windows, cooling summer breezes, and other factors to help reduce the cooling load of a building.

Passive solar system: Design of a building around the collection, storage, and distribution of solar resources.

Payback period: The amount of time required for a capital investment to pay back its initial investment, taking into account operating costs as well as profits earned.

Pedestrian pocket: A combination of housing, retail, and office space, placed within 1/4 mile of public transit, smaller in scale than planned unit developments or new towns.

Pedestrian scale: Urban designs that are oriented for pedestrians, making walking a safe, convenient, as well as interesting mode of travel.

Permeable: Allowing the passage of gases or fluids; relevant to moisture control and building materials.

Photocells: Light-sensitive cells that are used to activate switches at dawn and dusk.

Plasticizers: Chemicals that preserve the flexibility of soft plastics. Over time these agents off-gas, leaving the plastic brittle.

Porous paving: Paving materials that allow the infiltration of storm water, reducing the amount of runoff.

Post-consumer recycled content: Used materials that are separated from the waste stream for recycling.

Power: Expressed in watts (W), power is the rate at which energy is produced or consumed.

Pressure-treated wood: Wood chemically treated to prevent moisture and decay. The chemicals used can be a health hazard and must be treated and disposed of properly. Pressure-treated wood may create toxic fumes when burned.

Public transportation: Mass-transit systems, including buses and light rails. Sustainable building strategies place building sites near public transportation to promote commuting without using single-occupancy vehicles.

PVs (photovoltaics): Solid-state cells, typically made of silicon, that convert the sun's energy into electricity.

Radiant barrier: Material with a low emissivity rating, which is used to block the transfer of radiant heat across a space.

Radiant energy: Energy that radiates in all directions from its source in the form of electromagnetic waves.

Radiation: The transfer of heat through space by the straight-line passage of electromagnetic waves from a warm object to a cooler one.

Radon gas: An odorless, colorless gas that is radioactive and naturally exists in soil or rocks. When concentrations of radon build up inside a building, it can become a serious health hazard.

Rafter: A structural part that generally supports roofing or decking.

Rammed earth: A high-mass mixture of water, earth, and cement, used to build walls.

Recycled material: Material that has been withheld from disposal and reintroduced as feed-stock, which is then manufactured into marketable end products.

Refrigerant: A substance used as the working fluid in a cooling system.

Relative humidity: The percentage of water vapor present in the air in relation to the capacity of air to hold water vapor before condensing to liquid form.

Relite: Translucent panels or windows placed above doors or high up on walls for the purpose of allowing light to penetrate deeper into a building.

Renewable energy: Energy sources that are virtually inexhaustible or regenerative, such as solar radiation, biomass, wind, water, and geothermal heat.

Renewable resources: Resources that, when managed properly, can be produced as quickly as they are consumed.

Renovation: Upgrading an existing building with new materials while maintaining the building's original appearance.

Resource conservation: Methods that attempt to protect, preserve, or renew natural resources in ways that provide the best economic and social benefits.

Restoration: Returning a landscape or structure to its original design.

Retrofit: Upgrading, replacing, or improving a structure or piece of equipment within an existing facility.

Reuse: The use of a product or component of municipal solid waste more than once while maintaining the same original form. Reuse of materials helps to reduce the strain on renewable and nonrenewable resources alike.

Ridge: The peak of a sloped roof.

Risk assessment: The evaluation of risks posed to human health and/or the environment due to actual or potential presence or exposure of specific pollutants.

Runoff: Water that flows over land instead of absorbing into the soil.

R-value: A unit used to rate the thermal resistance of insulating materials. A material with a high R-value has more insulating properties than one with a lower value.

Salvage: Materials that have been diverted from the waste stream for the purpose of being reused.

SBS (sick-building syndrome): A condition that is defined by the symptoms that people show when they are in an unhealthy building. The symptoms include dizziness, irritated eyes, headaches, nausea, throat irritation, and coughing and usually stop when the person exits the building.

Sealant: A compound used to seal or secure something to prevent the leakage of air or moisture.

Sediment basin: A depression made in the ground and placed so as to hold sediment and debris on site.

Shading coefficient: Measured in intervals between zero and one, the ratio of the solar heat gain of a given material to that of a 1/8-inch clear double-strength glass. A lower-rated window transmits less solar heat.

SIP (structural insulated panel): Manufactured wood panels that contain a core of polystyrene, making them more resistant to air infiltration.

Sisal: A fibrous product derived from the sisal plant, used for floor coverings.

Site assessment: Conducted in the planning stages of development, the assessment of site characteristics such as soil, hydrology, typography, wetlands, wind direction, solar orientation, and existing habitats as well as community connectedness.

Site development costs: All costs included in the preparation of land for development, including demolition of existing structures, site preparation, and on- and off-site improvements.

Skylight: A glazed roof aperture used to allow daylight into a building.

Sludge: The sediment extracted from wastewater.

Soffit: An eave with a closed underside.

Solar access: Laying out a building and its landscaping in order to allow maximum availability to the sun's energy.

Solar collector: Any device that uses the sun's energy to provide energy for uses that would usually be supplied by a non-renewable energy source. Photovoltaic panels are a solar collector.

Solar energy: Energy received from waves of electromagnetic radiation that originate from the sun.

Source reduction: The practice of designing, manufacturing, and purchasing materials that promote resource conservation while reducing the amount of toxic waste being released into the waste stream and the environment.

Spec house: A home constructed with the speculation of finding a buyer.

Specifications: As provided in conjunction with blueprints or plans, detailed instructions that give necessary instructions that are otherwise not included in a plan such as: building materials, dimensions, colors, or special construction techniques.

Sprawl: The extension of shopping centers, small industries, and residential areas outside the boundaries of a city.

Straw-bale construction: An alternative building method that utilizes bales of straw for wall construction. This method reduces the need for lumber products, reuses an agricultural waste product, and achieves high insulation values.

Stud: Vertical structural part used to frame walls, generally wood or metal.

Sulphur dioxide: A byproduct of coal combustion, a colorless, irritating gas that contributes to acid rain.

Sunshades: Blocking devices used to prevent unwanted solar gain.

Superwindows: With R-values of 4.5 or higher, double or triple-glazed windows that contain an interior layer of Mylar-coated film.

Supply air: Typically a conditioned mixture of return air and outdoor air; the entire amount of air supplied to a building or space for ventilation purposes.

Sustainable: The ability to meet the needs of present generations without compromising the needs of future generations. For a human community to remain sustainable, it must not compromise biodiversity, must reuse or recycle as many materials as possible, must not consume resources faster than they are renewed, and must rely mainly on local or regional resources.

Sustainably sourced materials: Materials that have been manufactured and handled in a way that emphasizes the appropriate and efficient use of natural resources.

Thermal break: A material with low conductance used to reduce the flow of heat between two elements.

Thermal bridging: An element with high conductivity that may compromise the insulating value of the building envelope.

Thermal chimney: An area of a building used to control and utilize hot-air currents to help stimulate the flow of fresh air into a building.

Thermal conductance: The ability of a material to conduct heat.

Thermal storage capacity: The capacity of a building material to internally store heat from the sun.

Tipping fees: Fees charged for dumping large amounts of waste into a landfill.

TND (traditional neighborhood development): One of the main forms of new urbanism, a development that has an identifiable edge and a center that includes public space as well as commercial enterprise, encompasses a variety of activities and housing types, uses a modified grid system of interconnected streets and blocks, and gives high priority to public spaces.

Topography: The configuration of the surface and physical features of a place.

Topsoil: The top layer of soil containing more organic material than other layers of soil.

Transit-oriented development: The development of a mixed-use community within a 2,000-foot average walking distance from a transit stop. Such developments combine a core commercial area as well as residential, retail, office, and open spaces into a public domain that is easily accessible by foot, bike, or transit.

Trombe wall: A wall of masonry, oriented facing south, that has a layer of glass spaced a few inches away. Rays from the sun pass through the glass and are transformed into heat on the surface of the wall; the heat either passes into the building interior or is thermosyphoned through vents to interior spaces.

Truth window (or wall): A section of window or wall that is cut away to show its internal components.

Ultraviolet radiation: Electromagnetic radiation that has wavelengths between 4 and 400 nanometers, usually comes from the sun, and can pose a health risk that may lead to cancer or cataracts.

Urea-formaldehyde foam insulation: An insulating material once commonly found in crawl spaces and attics, emissions from which have been found to be a health hazard.

U-value: The measure of heat conductivity in or out of a substance when the temperature on one side is one degree different from the other. U-values are used to measure the performance of a window assembly or glazing.

Vapor: The gaseous form of a compound that is usually in a liquid or solid form.

Vapor retarder/vapor barrier: A material used to stop or reduce the seepage of water.

Variance: A special allowance to use a specific property or structure in a way that would not normally be allowed because of existing zoning laws.

VOC (volatile organic compound): A chemical compound known to cause nausea, tremors, and headaches, and believed to cause long-lasting harm. VOCs are commonly emitted from oil-based paints, solvent-based finishes, and other construction materials.

Warm-edge technology: The placement of low-conductance spacers near the edge of insulated glazing to reduce heat transfer.

Wastewater: Water that has been used for residential, farming, or industrial processes and contains dissolved or particulate contaminants.

Water budget: An estimation of the water needs of a facility, taking into account fixture and appliance flow rates, occupancy needs, and landscaping needs.

Water harvesting: The collection of runoff and rainwater to be used for various tasks such as irrigation and water features.

Watershed: An area of land that collects draining water due to topographical features.

Watt (W): A measure of electrical power. One watt is equivalent to one joule of work per second.

Wetland: The transitional land between terrestrial and aquatic ecosystems that is covered by water for part of the year. Wetlands serve as natural flood protection, naturally improve water quality, and are also important for their habitat and diversity of species.

Whole-systems thinking: Taking into consideration all the interconnected systems in order to find, address, and solve multiple problems at once.

Wingwall: An outside wall, attached perpendicular to exterior walls, stimulating air movement near the window for ventilation purposes.

Work: The use of force through a distance. Power defines the rate at which work is done, and energy is stored work. The basic unit of work is the joule, which is defined as the amount of work done while exerting one newton (N) over one meter.

Xeriscaping: Landscape design that emphasizes water conservation by using drought-resistant and drought-tolerant plants.

Zoning: Local government rulings that serve to prevent conflicts in land use while maintaining structure in the development and regulation of privately owned land.

APPENDIX TWO

Time-Saving Tips and Tables for Land Developers

Appendix 2

Land Use Restriction Addendum
Property Disposition Program
Officer Next Door/Teacher Next Door Sales Program

U.S. Department of Housing and Urban Development
Office of Housing
Federal Housing Commissioner

OMB Approval No. 2502-0306
(exp. 6/30/2004)

Public reporting burden for this collection of information is estimated to average 30 minutes per response, including the time for reviewing instructions, searching existing data sources, gathering and maintaining the data needed, and completing and reviewing the collection of information. This information is required to obtain benefits. HUD may not collect this information, and you are not required to complete this form, unless it displays a currently valid OMB control number.

This information is required in order to administer the Property Disposition Sales Program (24 CFR Part 291. The collection of information is required in order to provide a binding contract between the property purchaser and

HUD. A real estate broker or one of its agents completes this form. If this information were not collected, HUD would not be able to administer the Property Disposition Sales Program properly to avoid waste, mismanagement, and abuse. While no assurances of confidentiality are pledged to respondents, HUD generally discloses this data only in response to a Freedom of Information request. Failure to provide this information could affect your participation in HUD's Property Disposition Program.

Warning: Falsifying information on this or any other form of the Department is a felony. It is punishable by a fine not to exceed $250,000 and/or a prison sentence of not more than two years.

This Addendum is incorporated by reference to the FHA Sales Contract for the property located at

Property address _____

executed on this same day of _____, (date)(mm/dd/yyyy) between

Purchaser _____ and
Seller the Department of Housing and Urban Development.

1. Unless an exception is granted in writing by the Seller:
    a. The Purchaser is required to assign the sales contract (with Seller's approval) before or at the time of closing or participate in a three party closing with the Law Enforcement Officer or Teacher who will own and use the property as their sole residence for at least three years. Should the contract not be assigned prior to sales closing, the Purchaser must obtain certification from the Law Enforcement Officer's or Teacher's employer that the employee meets HUD's Law Enforcement Officer or Teacher definition and is in good standing with the employer and a certification from the Law Enforcement Officer or Teacher that he/she will own and use the property as their sole residence and not own any other residential real property for at least three years. The teacher must also certify that the home he/she intends to purchase is located in a school district / jurisdiction serviced by his/her employer. It is intended that the discount received by the Purchaser be passed on to the Law Enforcement Officer or Teacher.
    b. The Purchaser shall not resell the property for an amount in excess of 110 percent of the net development cost. Net development cost is the total cost of the project, including items such as acquisition cost, architectural fees, permits and survey expenses, insurance, rehabilitation, and taxes. It does not include a developer's fee. Total costs incurred by the Purchaser, including those for acquisition financing, management, rehabilitation, and selling expenses, are expected to be reasonable and customary for the area in which this property is located.
    c. The developer's fee provides for Purchaser's overhead and staffing costs related to the project, and may not exceed 10 percent of the net development cost.
    d. The property may not be occupied by or resold to any of the Purchaser's officers, directors, elected or appointed officials, business associates, or to any individual who is related by blood, marriage, or law to any of the above.
    e. There may be no conflict of interest with individuals or firms that may provide acquisition or rehabilitation funding, management or sales services, or other services associated with the project.
    f. A second mortgage and note apply to the resale to Law Enforcement Officers or Teachers. The second mortgage and note must be executed by the Law Enforcement Officer or Teacher. The appropriate documents have been received from HUD.
2. Purchaser must provide periodic reports, as specified by 24 CFR 291.210 and in the format and frequency specified by HUD, regarding the purchase and resale of properties subject to this Addendum.
3. This Addendum survives the expiration, if any, by operation of law or otherwise, of the FHA Sales Contract, and shall terminate five years from the date contained herein.

Purchaser's
Signature and Date (mm/dd/yyyy) _____

Witness's
Signature and Date (mm/dd/yyyy) _____

Seller Secretary of Housing and Urban Development

Type Name By _____

Signature and Date (mm/dd/yyyy) _____

Previous edition is obsolete.   ref. Handbook 4310.5   form **HUD-9548-B** (9/1999)

FIGURE A2.1 Sample land use restriction addendum.

Surveyor's Report

U.S. Department of Housing and Urban Development
Office of Housing
Federal Housing Commissioner

Instructions: Submit a completed, signed Surveyor's Report with all survey map/plat submissions.
See the Surveyor's Instructions for required map/plat submissions.
Identify pertinent observed and otherwise known conditions on the Surveyor's Report.

I certify that, on (date) _____, I made a survey of the premises standing in the name of _____

situated at (city, county, state): _____

known as street numbers _____

and shown on the accompanying survey entitled: _____

I made a careful inspection of said premises and of the buildings located thereon at the time of making such survey, and again, on (date) _____, and on such latter inspection, I found said premises to be standing in the name of: _____

In my professional opinion, the following information reflects the conditions observed on the date of the last site inspection or disclosed in the process of researching title to the premise, and I further certify that such conditions(s) are shown on the survey map/plat dated _____ or has/have been updated thereon under Revision Date _____.

1. Rights of way, old highways or abandoned roads, lanes or driveways, drains, sewer or water pipes over and across said premises:

2. Springs, streams, rivers, ponds or lakes located, bordering on or running through said premises:

3. Cemeteries or family burying grounds located on said premises:

4. Electricity, or electromagnetic/communications signal, towers, antenna, lines, or line supports located on, overhanging or crossing said premises:

5. Disputed boundaries or encroachments. (If the buildings, projections or cornices thereof or signs affixed thereto, fences or other indications of occupancy encroach upon adjoining properties or the like encroach upon surveyed premises, specify all such):

6. Earth moving work, building construction, or building additions within recent months:

7. Building or possession lines. (In case of city or town property specify definitely as to whether or not walls are independent walls or party walls and as to all easements of support or "Beam Rights." In case of country property report specifically how boundary lines are evidenced, that is, whether by fences or otherwise):

8. Recent street or sidewalk construction and/or any change in street lines either completed or proposed by and available from the controlling jurisdiction:

9. Flood hazard.

10. Site used as a solid waste dump, sump, or sanitary landfill.

Surveyor's Name: (print or type)	License Number:	Signature
		X

Previous editions are obsolete   form HUD-2457 (08/2003)
ref. Handbooks 4430.1 & 4460.1

FIGURE A2.2 Sample surveyor's report.

HUD Survey Instructions and Report
for Insured Multifamily Projects

U.S. Department of Housing and Urban Development
Office of Housing
Federal Housing Commissioner

OMB Approval No. 2502-0010 (exp. 06/30/2005)

Public reporting burden for this collection of information is estimated to average 0.5 hour per response, including the time for reviewing instructions, searching existing data sources, gathering and maintaining the data needed, and completing and reviewing the collection of information. This agency may not collect this information, and you are not required to complete this form, unless it displays a currently valid OMB control number.

This information is necessary to secure a marketable title and title insurance for the property that provides security for project mortgage insurance furnished under the FHA multifamily programs. This information assists in making determinations regarding the property's compliance with applicable program regulations, e.g., those pertaining to flood hazard, and in reaching underwriting determinations regarding property suitability and worth for the intended use. This information is mandatory. HUD does not assure confidentiality and there are no sensitive questions.

This survey is to be used in a loan transaction for which the U.S. Department of Housing and Urban Development (HUD) is to insure a multifamily project mortgage.

Its uses will include:
- [ ] Land title recordation (all cases).
- [ ] Site grading plan preparation (item 1 below).
- [ ] Plot plan design/redesign (item 2 below).

Special Project Features:
- [ ] Care Facility,
- [ ] Condo/Air-rights, and/or
- [ ] Other: (Specify)

Standards of Performance: In every instance the survey and map(s) and/or plat(s) must be made in accordance with the requirements for an "Urban Survey" and in compliance with the:

- Minimum Standard Detail Requirements and Classifications for ALTA/ACSM Land Title Surveys, as adopted by the American Land Title Association and American Congress on Surveying and Mapping, dated 1999;
- Table A, Optional Survey Responsibilities and Specifications, thereof, items 1 through 4 and 7 through 13 except for subitems 7b and 7c;
- and the following requirements as applicable:

1. **Site Grading Involved:** Comply with table A, item 5. Contours may not exceed 1-foot vertical intervals, except that 2-foot and 5-foot vertical intervals may be used where the mean site gradient exceeds 5 percent and 10 percent respectively. Where curbs and/or gutters exist, show top of curb and flow line elevations.

2. **Plot Plan Design/Redesign Involved:** Comply with Table A, Item 6.

3. **Condo/Air-rights Involved:** The surveyor must provide a survey made in accordance with any applicable jurisdictional requirements or, in the absence of such requirements, professionally recognized standards.

4. **Flood Hazard Involved:** Where any portion of the site is subject to flood hazard, show the 100 year return frequency flood hazard elevation and flood zone for all projects plus the 500 year return frequency flood hazard elevation and flood zone for care facility projects. For existing projects show the site elevation at the entrances, lowest habitable finish floor, and basement for each primary building. Take return frequency flood hazard elevations from the applicable Federal Flood Insurance Rate Map. Where such is not available, take the elevations from available State or local equivalent data, or when not available, work in conjunction with owner's engineer.

5. **Blanket Easement Involved.** Show on the map/plat the location of any facility that is located within or traverses the property under provisions of a blanket easement.

Additional Owner Requirements: The following requirements are not intended to void any other part of this instruction.

_____
_____
_____

Owner's Representative / Contact:
Name & Phone No: _____
Address: _____

Surveyor's Report: A current Surveyor's Report (not more than 120 days old) must be included with the survey map(s)/plat(s) submitted to HUD for: project design review, construction contract document sets, as required during construction, upon project completion; and with the map(s)/plat(s) used at initial and final closing.

Certification: The survey map/plat must bear the following certification:

"I hereby certify to the U.S. Department of Housing and Urban Development (HUD), *(Borrower), (Sponsor), (Lender), (Title Insurance Underwriter), (Other)*, and to their successors and assigns, that:

I made an on the ground survey per record description of the land shown hereon located in *(city or town, county, township, etc.)*, on *(date)*; and that it and this (these) map(s) was (were) made in accordance with the HUD Survey Instructions and Report, form HUD-2457, and the requirements for a Land Title Survey, as defined in the *Minimum Standard Detail Requirements for ALTA/ACSM Land Title Surveys* dated 1999.

To the best of my knowledge, belief and information, except as shown hereon: There are no encroachments either way across property lines; title lines and lines of actual possession are the same; and the premises are free of any (subject to a) 100/500 year return frequency flood hazard, and such flood free (flood) condition is shown on the Federal Flood Insurance Rate Map, Community Panel No: *(state, if none)*."

Previous editions are obsolete Page 1 of 2 form **HUD-2457** (08/2003)
ref. Handbooks 4430.1 & 4460.1

FIGURE A2.3 Sample HUD survey instructions and report.

Form RD 442-20
(Rev. 10-96)

UNITED STATES DEPARTMENT OF AGRICULTURE
RURAL DEVELOPMENT

FORM APPROVED
OMB NO. 0575-0015

RIGHT-OF-WAY EASEMENT

KNOW ALL MEN BY THESE PRESENTS:

That in consideration of One Dollar ($1.00) and other good and valuable consideration paid to _____ and _____,

hereinafter referred to as GRANTOR, by _____,
hereinafter referred to as GRANTEE, the receipt of which is hereby acknowledged, the GRANTOR does hereby grant, bargain, sell, transfer, and convey unto the GRANTEE, its successor and assigns, a perpetual easement with the right to erect, construct, install, and lay, and thereafter use, operate, inspect, repair, maintain, replace, and remove

_____

_____

_____

over, across, and through the land of the GRANTOR situate in _____ County,
State of _____, said land being described as follows:

_____

_____

_____

together with the right of ingress and egress over the adjacent lands of the GRANTOR, his successors and assigns, for the purposes of this easement.

The easement shall be _____ feet in width, the center line of which is described as follows:

_____

_____

_____

The consideration hereinabove recited shall constitute payment in full for any damages to the land of the GRANTOR, his successors and assigns, by reason of the installation, operation, and maintenance of the structures or improvements referred to herein. The GRANTEE covenants to maintain the easement in good repair so that no unreasonable damage will result from its use to the adjacent land of the GRANTOR, his successors and assigns.

The grant and other provisions of this easement shall constitute a covenant running with the land for the benefit of the GRANTEE, its successors and assigns.

IN WITNESS WHEREOF, the GRANTORS have executed this instrument this _____ day of _____

19 _____:

_____ (SEAL)

_____ (SEAL)

Public reporting burden for this collection of information is estimated to average 1 hour per response, including the time for reviewing instructions, searching existing data sources, gathering and maintaining the data needed, and completing and reviewing the collection of information. Send comments regarding this burden estimate or any other aspect of this collection of information, including suggestions for reducing this burden, to U.S. Department of Agriculture, Clearance Officer, STOP 7602, 1400 Independence Avenue, S.W., Washington, D.C. 20250-7602. **Please DO NOT RETURN this form to this address.** Forward to the local USDA office only. You are not required to respond to this collection of information unless it displays a currently valid OMB control number.

RD 442-20 (Rev. 10-96)

FIGURE A2.4 Sample right-of-way easement form.

Appendix 2

PERFORMANCE BOND FOR OTHER THAN CONSTRUCTION CONTRACTS (See instructions on reverse)	DATE BOND EXECUTED (Must be same or later than date of contract)	OMB No.: 9000-0045

Public reporting burden for this collection of information is estimated to average 25 minutes per response, including the time for reviewing instructions, searching existing data sources, gathering and maintaining the data needed, and completing and reviewing the collection of information. Send comments regarding this burden estimate or any other aspect of this collection of information, including suggestions for reducing this burden, to the FAR Secretariat (MVR), Federal Acquisition Policy Division, GSA, Washington, DC 20405

PRINCIPAL (Legal name and business address)	TYPE OF ORGANIZATION ("X" one)
	☐ INDIVIDUAL ☐ PARTNERSHIP
	☐ JOINT VENTURE ☐ CORPORATION
	STATE OF INCORPORATION

SURETY(IES) (Name(s) and business address(es))	PENAL SUM OF BOND			
	MILLION(S)	THOUSAND(S)	HUNDRED(S)	CENTS
	CONTRACT DATE		CONTRACT NO.	
	OPTION DATE		OPTION NO.	

OBLIGATION:

We, the Principal and Surety(ies), are firmly bound to the United States of America (hereinafter called the Government) in the above penal sum. For payment of the penal sum, we bind ourselves, our heirs, executors, administrators, and successors, jointly and severally. However, where the Sureties are corporations acting as co-sureties, we, the Sureties, bind ourselves in such sum "jointly and severally" as well as "severally" only for the purpose of allowing a joint action or actions against any or all of us. For all other purposes, each Surety binds itself, jointly and severally with the Principal, for the payment of the sum shown opposite the name of the Surety. If no limit of liability is indicated, the limit of liability is the full amount of the penal sum.

CONDITIONS:

The Principal has entered into the contract identified above.

THEREFORE:

The above obligation is void if the Principal: (1) Performs and fulfills all the undertakings, covenants, terms, conditions, and agreements of the contract during either the base term or an optional term of the contract and any extensions thereof that are granted by the Government, with or without notice to the Surety(ies), and during the life of any guaranty required under the contract, and (2) performs and fulfills all the undertakings, covenants, terms, conditions, and agreements of any and all duly authorized modifications of the contract that hereafter are made. Notice of those modifications to the Surety(ies) is waived.

The guaranty for a base term covers the initial period of performance of the contract and any extensions thereof excluding any options. The guaranty for an option term covers the period of performance for the option being exercised and any extensions thereof.

The failure of a surety to renew a bond for any option term shall not result in a default of any bond previously furnished covering any base or option term.

WITNESS:

The Principal and Surety(ies) executed this performance bond and affixed their seals on the above date.

PRINCIPAL			
SIGNATURE(S)	1. (Seal)	2. (Seal)	Corporate Seal
NAME(S) & TITLE(S) (Typed)	1.	2.	

INDIVIDUAL SURETY(IES)		
SIGNATURE(S)	1. (Seal)	2. (Seal)
NAME(S) (Typed)	1.	2.

CORPORATE SURETY(IES)				
SURETY A	NAME & ADDRESS		STATE OF INC.	LIABILITY LIMIT $
	SIGNATURE(S)	1.	2.	Corporate Seal
	NAME(S) & TITLE(S) (Typed)	1.	2.	

AUTHORIZED FOR LOCAL REPRODUCTION
Previous edition not usable

STANDARD FORM 1418 (REV. 2-99)
Prescribed by GSA-FAR (48 CFR) 53.228(b)

FIGURE A2.5 Sample performance bond for other than construction projects. (continued)

SURETY					STATE OF INC.	LIABILITY LIMIT	
SURETY B	NAME & ADDRESS					$	Corporate Seal
	SIGNATURE(S)	1.	2.				
	NAME(S) & TITLE(S) *(Typed)*	1.	2.				
SURETY C	NAME & ADDRESS					$	Corporate Seal
	SIGNATURE(S)	1.	2.				
	NAME(S) & TITLE(S) *(Typed)*	1.	2.				
SURETY D	NAME & ADDRESS					$	Corporate Seal
	SIGNATURE(S)	1.	2.				
	NAME(S) & TITLE(S) *(Typed)*	1.	2.				
SURETY E	NAME & ADDRESS					$	Corporate Seal
	SIGNATURE(S)	1.	2.				
	NAME(S) & TITLE(S) *(Typed)*	1.	2.				
SURETY F	NAME & ADDRESS					$	Corporate Seal
	SIGNATURE(S)	1.	2.				
	NAME(S) & TITLE(S) *(Typed)*	1.	2.				
SURETY G	NAME & ADDRESS					$	Corporate Seal
	SIGNATURE(S)	1.	2.				
	NAME(S) & TITLE(S) *(Typed)*	1.	2.				

BOND PREMIUM	RATE PER THOUSAND ($)	TOTAL ($)

INSTRUCTIONS

1. This form is authorized for use in connection with Government contracts. Any deviation from this form will require the written approval of the Administrator of General Services.

2. Insert the full legal name and business address of the Principal in the space designated "Principal" on the face of the form. An authorized person shall sign the bond. Any person signing in a representative capacity (e.g., an attorney-in-fact) must furnish evidence of authority if that representative is not a member of the firm, partnership, or joint venture, or an officer of the coporation involved.

3. (a) Corporations executing the bond as sureties must appear on the Department of the Treasury's list of approved sureties and must act within the limitation listed therein. Where more than one corporate surety is involved, their names and addresses shall appear in the spaces (Surety A, Surety B, etc.) headed "CORPORATE SURETY(IES)." In the space designated "SURETY(IES)" on the face of the form, insert only the letter identification of the sureties.

(b) Where individual sureties are involved, a completed Affidavit of Individual Surety (Standard Form 28) for each individual surety, shall accompany the bond. The Government may require the surety to furnish additional substantiating information concerning their financial capability.

4. Corporations executing the bond shall affix their corporate seals. Individuals shall execute the bond opposite the word "Corporate Seal", and shall affix an adhesive seal if executed in Maine, New Hampshire, or any other jurisdiction requiring adhesive seals.

5. Type the name and title of each person signing this bond in the space provided.

6. Unless otherwise specified, the bond shall be submitted to the contracting office that awarded the contract.

STANDARD FORM 1418 (REV.2-99) **BACK**

FIGURE A2.5 Sample performance bond for other than construction projects.

PERFORMANCE AGREEMENT
of Allegany County, Maryland

KNOW ALL MEN BY THESE PRESENTS: that _____. at _____, hereinafter called the Principal, and _____, hereinafter called the Surety, are held and firmly bound unto The Allegany County Commissioners, Allegany Office Complex, 701 Kelly Road, Cumberland, Md. 21502, hereinafter called the County, in the total aggregate penal sum of _____ dollars($_____.00) lawful money of the United States, for the payment of which sum well and truly to be made, we bind ourselves, our heirs, executors, administrators, successors, and assigns, jointly and severally, firmly by these presents.

THE OBLIGATION is such that the Principal must construct, implement, and make fully functional Stormwater Management Plan, File# ____SS____, approved the _____ day of _____, 200__, being a condition of Land Use Permit(LUP) Number _____.

NOW, THEREFORE, if the Principal shall well, truly and faithfully perform its duties, all the undertakings, covenants, terms, conditions, and agreements of said Permit during the original term thereof, and any extensions thereof which may be granted by the County, with or without notice to the Surety and during the guaranty period and if the Principal shall satisfy all claims and demands incurred under LUP# _____ and shall fully indemnify and save harmless the County from all costs and damages which it may suffer by reason of failure to do so, and shall reimburse and repay the County all outlay and expenses which the County may incur in making good any default, then this obligation shall be void, otherwise to remain in full effect.

PROVIDED, FURTHER, that the liability of the PRINCIPAL AND SURETY hereunder to the County shall be subject to the same limitations and defenses as may be available to them against a claim hereunder by the County, provided, however, that the County may, in its option, perform any obligations of the County required by the Permit and Allegany County Land Use Regulations.

PROVIDED, FURTHER, that the said Surety, for value received hereby stipulates and agrees that no change, extension of time, alteration, or addition to the terms of the Permit or Plan be performed thereunder or the Specifications accompanying same shall in any way affect its obligation to this Agreement, and it does hereby waive notice of any such change, extension of time, alteration, or addition to the terms of the Permit or Plan.

PROVIDED, FURTHER, that is expressly agreed that the Agreement shall be deemed amended automatically and immediately, without formal and separate amendments hereto, upon amendment to the Permit or Plan, so as to bind the principal and the Surety to the full and faithful performance of the Permit or Plan as so amended. The term "Amendment", as used in this Agreement, and whether referring to this Agreement, the Plan or Permit, shall include any alteration, addition extension, or modification of any character whatsoever.

PROVIDED, FURTHER that no final settlement between the County and the Principal shall occur until he County accepts an ASCD approved As-Built Certification prepared in accordance with the Allegany County Stormwater Management Ordinance.

PROVIDED, FURTHER that the County is the only beneficiary of this Agreement.

FIGURE A2.6 Sample performance agreement. (continued)

Performance Agreement Cont'd
MM/DD/YY
Page 2 of 2

    IN WITNESS THEREOF, this instrument is executed in two counterparts, each one of which shall be deemed an original, this the _____th day of _____.

ATTEST:

_____
Principal

_____
Secretary

NOTARY: STATE OF _____, County of _____, TO WIT:

    I hereby certify that on this _____ day of _____, 200__, before me the subscriber, a Notary Public of the State aforesaid, personally appeared _____ _____, known to me, or satisfactorily proven to be the person(s) whose name is subscribed to the foregoing instrument, who acknowledged that he has executed it for the purposes therein set forth, and that it is his act and deed.

    In witness thereof, I have set my hand and Notarial Seal, the day and year first written above.

My Commission expires on _____

_____
Notary Public

ATTEST:

_____
Surety

_____
Attorney in Fact

NOTARY: STATE OF _____, County of _____, TO WIT:

    I hereby certify that on this _____ day of _____, 200__, before me the subscriber, a Notary Public of the State aforesaid, personally appeared _____ _____, known to me, or satisfactorily proven to be the person(s) whose name is subscribed to the foregoing instrument, who acknowledged that he has executed it for the purposes therein set forth, and that it is his act and deed.

    In witness whereof, I have set my hand and Notarial Seal, the day and year first written above.

_____
Notary Public

My Commission expires on _____

\\forms\swm\swmbond.doc
rev 9/02   [E4650/LDS000]

FIGURE A2.6   Sample performance agreement.

Appendix 2

Haywood County Erosion Control Program
Haywood County Annex II, 1233 North Main Street
Waynesville, N.C. 28786   Phone: 828-452-6706

LAND DISTURBING SURETY BOND

Consult instructions for completion.
PRINCIPAL INFORMATION:

Name:_____

D.B.A.:_____

Site or Project Name:_____

Land Owner(s) of Record:_____

Address:_____

City/State/zip:_____

Phones and Fax:_____

E-Mail Address:_____

State License or Registration # s:_____

SURETY INFORMATION:

Name:_____

Address:_____

City/State/Zip:_____

Phones and Fax:_____

E-Mail Address:_____

Bond#:_____

State of North Carolina
County of Haywood

KNOW ALL MEN BY THESE PRESENTS THAT WE,_____,
                                                                                    (Principal Name)

as Principal and _____, as Surety, are held and firmly bound unto the County
                                 (Surety Name)
of Haywood, in the sum of ($_____.00)_____to the
payment where we bond ourselves our heirs, executors, administrators, and assigns, firmly by these present.

   WHEREAS, the above bounden Principal has applied for an **EROSION CONTROL PLAN APPROVAL AND LAND-DISTURBING PERMIT**, in Haywood County, North Carolina.

   The condition of this obligation is such that:

   WHEREAS, the said Principal is or desires to be engaged in a **LAND-DISTURBING ACTIVITY** within Haywood County on a parcel or tract of land which is known in the official registry of the Haywood County Land Records Office by the PROPERTY IDENTIFICATION NUMBER of:_____, and said parcel or tract may also be found in DEED BOOK:_____and PAGE #:_____in the Haywood County Register of Deeds Office, and

   WHEREAS, there have been promulgated by Haywood County, certain rules and regulations for the conduct of such land-disturbing activities as proposed by the Principal, and

(PAGE 1 OF 2, HAYWOOD COUNTY BOND FORM FOR LAND-DISTURBING ACTIVITY)

FIGURE A2.7   Sample land disturbing surety bond. (continued)

WHEREAS, specific to the conditions creating the requirement of this Surety Bond, the said land-disturbing activity is subject to /154.7 of Chapter 154 of the Haywood County Code of Ordinances: Erosion and Sediment Control, and

NOW THEREFORE, if the said Principal shall well and truly perform the land-disturbing activity from the time of undertaking to completion within the guidelines set forth in the approved erosion and sediment control plan for the project **and** Chapter 154 of Haywood County s Code of Ordinances (Erosion and Sediment Control), Haywood County will make no demand to redeem the bond. However, the said Principal and the said Surety shall well and truly pay to Haywood County all applicable surety bond funds stated herein if the land-disturbing activity in is non-compliance with said Ordinance for 90 working days after a Notice of Violation is received by the Principal.

It is expressly understood that this bond may be canceled by the Surety only at the expiration of thirty (30) calendar days from the date upon which the Surety shall have filed with the Haywood County Erosion Control Program **and** the Haywood County Finance Director written notice to so cancel. This provision however, shall not operate to relieve, release or discharge the Surety from any liability, civil penalties or criminal penalties already accrued or which shall accrue before the expiration of the thirty (30) day period. It is expressly understood that if the bond lapses or expires prematurely, the Land-Disturbing Permit will be revoked, and an application for a new Land-Disturbing Permit must then be submitted. It is expressly understood that upon forfeiture of applicable surety, the Principal does hereby grant to Haywood County the right to enter said property at reasonable times and perform work upon said property to the value extent of the bond and only for the purpose of installation of sufficient erosion and sediment control measures and devices on the site in accordance with Chapter 154 of the Haywood County Code of Ordinances: Erosion and Sediment Control. It is expressly understood that forfeited surety shall be also used to establish erosion control structures or ground covers in accordance with an approved sediment control plan.

This is the _____ day of _____.

PRINCIPAL: _____

WITNESS TO PRINCIPAL: _____

SURETY SEAL: _____

WITNESS TO SURETY: _____

ATTORNEY-IN-FACT (SURETY): _____

Details of Application:

1. The number of acres to be disturbed, including all borrow and waste areas and all access and haul roads will be stated as follows to the nearest tenth of an acre: _____.

2. Dollar amount (U.S.A.) per acre to be posted (fractions of acres will be prorated): $_____.

3. The total amount of the bond will now be stated as follows: $_____.

4. An original copy of all bond forms must be received by Haywood County in order for the bond to be considered valid and before the Land-Disturbing Permit may be issued.

(PAGE 2 OF 2, HAYWOOD COUNTY BOND FORM FOR LAND-DISTURBING ACTIVITY)

FIGURE A2.7 Sample land disturbing surety bond. (continued)

Haywood County Erosion Control Program
Building Box 12, Haywood County Annex II
1233 North Main Street
Waynesville, NC 28786

Phone: 828-452-6706 or Fax: 828-452-6767

INSTRUCTIONS FOR COMPLETING LAND-DISTURBING ACTIVITY SURETY BONDS

INFORMATION

In compliance with the Haywood County Erosion and Sediment Control Ordinance, application for a permit to disturb five or more acres shall require the posting of a surety bond with the County in the form of an escrow account, an account guaranteed by an established surety company or other instruments satisfactory to the County Attorney.

CASH BONDS can only be accepted with a certified check or money order and are not interest bearing.

FAXED or **PHOTOCOPIED** bonds will be accepted as evidence that the bond has been issued if the form is completed, signed and sealed. However the original must be received by the County before the Land-Disturbing Permit can be issued.

INSTRUCTIONS

1. **THE BOND MUST BE EXECUTED ON HAYWOOD COUNTY S BOND FORM** and completed by a Surety company or Cash Principal. Bonds may only be cancelled by a written 30 day notice to the County by the Principal or the Surety. However, if the bond is canceled before the site is issued a Certificate of Compliance by the County, the Land-Disturbing Permit may be revoked.

2. **IF THE PRINCIPAL IS A PARTNERSHIP** or other person engaging in business under an assumed name, a copy of the Certificate of Assumed Name must be attached to the bond form. At least one partner or person must sign their full legal name as the legal Partnership or Business Representative as Principle.

3. **IF THE PRINCIPAL IS A CORPORATION**, a Registered Agent must also sign their full legal name as Principal.

4. **MAILING ADDRESSES** including zip codes, office phone numbers, fax numbers and cell numbers must be included for Principal and Surety.

5. **LIST ALL STATE LICENSE NUMBERS** or professional registration numbers held by the Principal.

6. **BOND NUMBER** is to be assigned by the Surety company. If the bond number has not been assigned, please send rider or endorsement listing the assigned number immediately.

7. **BOND AMOUNT** will be determined by the County by multiplying the number of acres to be disturbed by an amount within the limits specified in the Ordinance. The required bond amount per acre will be fairly determined by The Erosion Control Office and will generally be based on difficulty of site stabilization upon forfeiture of applicable surety.

8. **SIGNATURES** of Principal and Attorney In Fact (Surety) **ARE REQUIRED.**

9. **INSURANCE COMPANY S CORPORATE SEAL** must be affixed on bond. **A Notary seal or Principal s corporate seal are not acceptable.**

FIGURE A2.7 Sample land disturbing surety bond.

City of Clearwater
Development Services Department
100 South Myrtle Avenue, Clearwater, FL 33756
Phone 727-562-4567 Fax 727-562-4576

Erosion Control
Revised September 2000

The <u>PURPOSE</u> of this notice is to inform the prime contractor that the City of Clearwater will hold them responsible for soil erosion from their site.

The City of Clearwater Planning and Development Services Department has the responsibility to minimize the amount of soil erosion into the City's streets, storm sewers, adjacent properties and waterways. The construction of a new residence or commercial sites and/or major remodeling of an existing site create many instances whereby soil erosion can occur. These instances are usually the result of contractors and subcontractors accessing the property with equipment and/or construction materials. Subsequent rainstorms redistribute the eroded soil into the adjacent streets, storm systems, adjacent properties and waterways.

When erosion takes place, the Land Resource Specialist inspector or a Public Works inspector will place a correction notice at the site. The procedure will be as follows:

- 1st Occurrence – Warning
- 2nd Occurrence - $32.00 reinspection fee
- 3rd Occurrence - $80.00 reinspection fee
- 4th Occurrence – Stop work order

Dependent on the severity of the erosion, the City's Public Works Department may elect to rectify the erosion problem and charge the contractor accordingly.

The attached drawings and details are recommendations for the contractor to use as means to support the site from eroding. The contractor may elect to shovel and sweep the street daily or on an as needed basis. However, erosion must be held in check.

If the contractor would like to meet with the Land Resource Specialist inspector or Public Works inspector on any particular site, please contact Ray Boler (Public Works) at 727-462-6585 ext.265 or Rick Albee (Land Resources) at 727-562-4741.

Erosion Control Required

City of Clearwater's Land Development Code Section 3-702 requires erosion control on all land development projects.

Erosion control must be in place and maintained throughout the job. Failure to do so may result in additional costs and time delays to the permit holder.

Contact the Land Resources Specialist with specific questions at 727-562-4741.

Erosion Control.doc

FIGURE A2.8 Sample erosion control form. (continued)

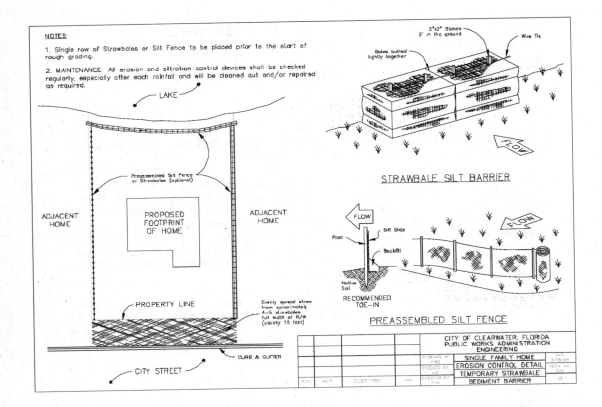

FIGURE A2.8 Sample erosion control form.

Allegany Soil Conservation District

11602 Bedford Road, NE   Cumberland, Maryland 21502   Phone 301-777-1747   Fax 301-777-7632

STANDARD EROSION AND SEDIMENT CONTROL PLAN FOR MINOR EARTH DISTURBANCES

Permit Number #_____ Date_____

Location_____

OWNER_____ Phone:_____

DEVELOPER or BUILDER_____ Phone:_____

Total lot area: _____ [ ] sf   **Total area disturbed**: _____ [ ] sf

_____ [ ] acres   _____ [ ] acres

Distance between disturbed area and any free-flowing stream_____ feet.

    I certify that I have the authority to make the foregoing application; that the information above and on the attached plot plan is correct; and that I agree to meet all the limitations and conditions set forth by this agreement.

OWNER'S SIGNATURE _____ Date _____

[ ] APPLICANT'S SIGNATURE
[ ] DEVELOPER'S SIGNATURE   Date _____
[ ] BUILDER'S SIGNATURE

COMPANY NAME_____

APPROVED:

    Allegany Soil Conservation District _____ Date _____

                       Expiration Date _____ 200__

PLAN CRITERIA

I. GENERAL

    This Standard Erosion and Sediment Control Plan may be used instead of a detailed plan for earth disturbances where all of the following conditions are met:

1. The owner, builder, or developer is not the same owner, builder, or developer of any contiguous lots undergoing development.

2. The undeveloped lot is completely vegetated.

3. No slopes steeper than 3 horizontal units to 1 vertical unit (33%) will be disturbed.

FIGURE A2.9   Sample erosion and sediment control form. (continued)

4. No more than 5,000 square feet or 250 cubic yards will be disturbed. Earth disturbances up to 20,000 square feet may be allowed for certain sites, pending Allegany SCD review.

5. Attached plot plan shows the proposed development, with arrows indicating the drainage pattern of the site.

II. CONDITIONS

1. The applicant shall notify the Allegany Soil Conservation District (301-777-1747, Ext. 109) at least 48 hours prior to commencing clearing or grading.

2. Access to the site and this plan shall be available at all times for inspection by the inspection agency.

3. In the event that the applicant fails to provide adequate sediment control according to the provisions of this plan, the inspection agency reserves the right to require corrective action.

4. Nothing herein relieves the applicant from complying with any and all other State or County regulations.

5. This Standard Erosion and Sediment control Plan will remain valid for 2 years from the approval date.

III. GRADING

1. Initial earth disturbance shall be limited to that necessary to install sediment control measures.

2. The permanent driveway or entrance location shall be used as a stabilized construction entrance. Two inch stone shall be placed at least 6 inches deep, 30 feet long, and 10 feet wide. The entrance shall be top dressed with stone as necessary to prevent tracking of sediment onto public streets or rights-of-way.

3. At any location where surface runoff from disturbed or graded areas flows off the property, silt fence shall be installed to prevent sediment from being transported off-site.

4. Swales or other areas that transport concentrated flow shall be sodded. Downspouts shall be protected by splashblocks or sod.

5. Grading shall not impair existing surface drainage, create an erosion hazard, or create a source of sediment to any adjacent watercourse of property.

6. Final graded slopes shall be no steeper than 3 horizontal units to 1 vertical unit (33%).

IV STABILIZATION

Following initial soil disturbance or redisturbance, permanent or temporary stabilization shall be completed within:

1. Seven calendar days for the surface of all perimeter controls, and all perimeter slopes.

2. Fourteen calendar days for all other disturbed or graded areas.

V. SPECIFICATIONS

For specifications regarding silt fence or straw bale dike installation and temporary and permanent stabilization practices, reference the "1994 Maryland Standards and Specifications for Soil Erosion and Sediment Control", or contact the Allegany Soil Conservation District.

FIGURE A2.9 Sample erosion and sediment control form. (continued)

E&SC/Zoning Site Plan
Page 3 of 3

STANDARD SITE PLAN

LUP#_____

from Part V of *Standard Erosion and Sediment Control Plan for Minor Earth Disturbances*.

SPECIFICATIONS: For specifications regarding silt fence or straw bale dike installation and temporary and permanent stabilization practices, reference the "1994 Maryland Standards and Specifications for Soil Erosion and Sediment Control", or contact the Allegany Soil Conservation District.

SCALE:_____:_____ [ ] No Scale   Zoning District_____   Election District_____
NOTE: Values noted reflect [ ] actual distance(s); [ ] minimum requirements in accordance with Zoning Ordinance.

LEGEND

Symbol	Description	Symbol	Description
⊕	well	---SS---/---W---	sewer line/water line
==========	road	▨▨▨	sewage disposal area
— - - - —	drainway	----- -- ------	property line
%↘	slope	- - -LOD- -	limits of disturbance(LOD)
-SF---SF---SF-	silt fence	Γ	rain leader w/ flow direction
[DW]	drywell	SCE	stabilized construction entrance(SCE)

REVISED

I hereby agree to comply with all regulations and codes, which are applicable hereto. I further agree that any misstatement or misrepresentation of facts presented as part of this application, or change to proposal without approval of the agencies concerned, shall constitute sufficient grounds for the disapproval or revocation of the subject permit. I hereby affirm that I own the property which is the subject of this application; or that I am the duly designated representative of the property owner, and that I possess the legal authority to make this *Affidavit* on behalf of myself or the owner for whom I am acting. I do solemnly declare and affirm under the penalties of perjury that the contents of this Application are true and correct to the best of my knowledge, information and belief.

Applicants Signature _____   Date _____
Approved By: _____
Bernie Connor, Sediment Control Planner   Date _____

\\forms\sec\ standard plan.doc
rev 9/02   [E4650]

TAKEN BY:_____

FIGURE A2.9   Sample erosion and sediment control form.

244 Appendix 2

PLACER COUNTY PLANNING DEPARTMENT *Reserved for Date Stamp*

AUBURN OFFICE TAHOE OFFICE
11414 B Avenue 565 W. Lake Blvd./P. O. Box 1909
Auburn, CA 95603 Tahoe City CA 96145
530-886-3000/FAX 530-886-3080 530-581-6280/FAX 530-581-6282
Website: www.placer.ca.gov/planning E-Mail : planning@placer.ca.gov

ENVIRONMENTAL IMPACT ASSESSMENT QUESTIONNAIRE

Receipt No. _____ Filing Fee: _____

Pursuant to the policy of the Board of Supervisors, the Planning Department cannot accept applications on tax delinquent property or property with existing County Code violations.

SEE FILING INSTRUCTIONS ON LAST PAGE OF THIS APPLICATION FORM

(ALL) 1. Project Name (same as on IPA) _____

PLNG 2. What is the general land use category for the project? (e.g.: residential, commercial, agricultural, or industrial, etc.) _____

PLNG 3. What is the number of units or gross floor area proposed? _____

DPW 4. Are there existing facilities on-site (buildings, wells, septic systems, parking, etc.)? Yes____ No____
If yes, show on site plan and describe: _____

DPW 5. Is adjacent property in common ownership? Yes____ No____ Acreage_____
Assessor's Parcel Numbers _____

PLNG 6. Describe previous land use(s) of site over the last 10 years: _____

GEOLOGY & SOILS

NOTE: Detailed topographic mapping and preliminary grading plans may be required following review of the information presented below.

DPW 7. Have you observed any building or soil settlement, landslides, slumps, faults, steep areas, rock falls, mud flows, avalanches or other natural hazards on this property or in the nearby surrounding area? Yes____ No____

DPW 8. How many cubic yards of material will be imported? _____ Exported? _____ Describe material sources or disposal sites, transport methods and haul routes: _____

DPW 9. What is the maximum proposed depth and slope of any excavation? _____
Fill? _____

DPW 10. Are retaining walls proposed? Yes____ No____. If yes, identify location, type, height, etc: _____

DPW 11. Would there be any blasting during construction? Yes____ No____ If yes, explain: _____

DPW 12. How much of the area is to be disturbed by grading activities? _____

PLNG 13. Would the project result in the direct or indirect discharge of sediment into any lakes or streams?
DEH Yes____ No____ If yes, explain: _____

DPW 14. Are there any known natural economic resources such as sand, gravel, building stone, road base rock, or mineral deposits on the property? Yes____ No____ If yes, describe: _____

FIGURE A2.10 Sample environmental impact assessment questionnaire. (continued)

NOTE: *Preliminary drainage studies may be required following review of the information presented below.*

DPW 15. Is there a body of water (lake, pond, stream, canal, etc.) within or on the boundaries of the property? Yes_____ No_____ If yes, name the body of water here and show location on site plan: _____
_____

DEH 16. If answer to #15 is yes, would water be diverted from this water body? Yes___ No___

DEH 17. If yes, does applicant have an appropriative or riparian water right? Yes____ No____

DEH 18. Where is the nearest off-site body of water such as a waterway, river, stream, pond, lake, canal, irrigation ditch, or year-round drainage-way? Include name, if applicable: does applicant have an appropriative or riparian water right? Yes____ No____

What percentage of the project site is presently covered by impervious surfaces? _____

After development? _____

DPW 19. Would any run-off of water from the project enter any off-site canal/stream? Yes_____ No_____
DEH   If answer is yes, identify: _____

DEH 20. Will there be discharge to surface water of waste waters other than storm water run-off? Yes_____ No_____

If yes, what materials will be present in the discharge? _____

What contaminants will be contained in storm water run-off? _____

DPW 21. Would the project result in the physical alteration of a body of water? Yes____ No____ If so, how? _____
_____

Will drainage from this project cause or exacerbate any downstream flooding condition? Yes_____ No_____ If yes, explain: _____

DPW 22. Are any of the areas of the property subject to flooding or inundation? Yes_____ No_____ If yes, accurately identify the location of the 100-year floodplain on the site plan.

DPW 23. Would the project alter drainage channels or patterns? Yes____ No____ If yes, explain: _____
DEH   _____

VEGETATION AND WILDLIFE

NOTE: *Detailed studies or exhibits such as tree surveys and wetland delineations may be required following review of the information presented below. Such studies or exhibits may also be included with submittal of this questionnaire. (See Filing Instructions #8 and #9 for further details.)*

PLNG 24. Describe vegetation on the site, including variations throughout the property: _____
_____

PLNG 25. Estimate how many trees of 6-inches diameter or larger would be removed by the ultimate development of this project as proposed: _____

PLNG 26. Estimate the percentage of existing trees which would be removed by the project as proposed _____

PLNG 27. What wildlife species are typically found in the area during each of the seasons? _____
_____

PLNG 28. Are rare or endangered species of plants or animals (as defined in Section 15380 of the California Environmental Quality Act Guidelines) found in the project area? _____

PLNG 29. Are any Federally listed threatened or endangered plants, or candidates for listing, present on the project site as proposed? If uncertain, a list is available in the Planning Department: _____

PLNG 30. Will the project as proposed displace any rare or endangered species (plants/animals)? _____

FIGURE A2.10 Sample environmental impact assessment questionnaire. (continued)

246   Appendix 2

PLNG 31. What changes to the existing animal communities' habitat and natural communities will the project cause as proposed? _____

PLNG 32. Is there any rare, natural community (as tracked by the California Department of Fish and Game Natural Diversity Data Base) present on the proposed project? _____

PLNG 33. Do wetlands or stream environment zones occur on the property (i.e., riparian, marsh, vernal pools, etc.)? Yes_____ No_____

PLNG 34. If yes, will wetlands be impacted or affected by development of the property? Yes_____ No_____

PLNG 35. Will a Corps of Engineers wetlands permit be required? Yes_____ No_____

PLNG 36. Is a letter from the U.S. Army Corps of Engineers regarding the wetlands attached? Yes_____ No_____

FIRE PROTECTION

DPW 37. How distant are the nearest fire protection facilities? _____
Describe: _____

DPW 38. What is the nearest emergency source of water for fire protection purposes? _____
Describe the source and location: _____

DPW 39. What additional fire hazard and fire protection service needs would the project create? _____
What facilities are proposed with this project? _____
For single access projects, what is the distance from the project to the nearest through road? _____
Are there off-site access limitations that might limit fire truck accessibility, i.e. steep grades, poor road alignment or surfacing, substandard bridges, etc.? Yes_____ No_____ If yes, describe: _____

NOISE

NOTE: *Project sites near a major source of noise, and projects which will result in increased noise, may require a detailed noise study prior to environmental determination.*

DEH 40. Is the project near a major source of noise? _____ If so, name the source(s): _____

DEH 41. What noise would result from this project - both during and after construction? _____

AIR QUALITY

NOTE: *Specific air quality studies may be required by the Placer County Air Pollution Control District (APCD). It is suggested that applicants with residential projects containing 20 or more units, industrial, or commercial projects contact the APCD before proceeding.*

APCD 42. Are there any sources of air pollution within the vicinity of the project? If so, name the source(s):_____

APCD 43. What are the type and quantity of vehicle and stationary source (e.g. woodstove emissions, etc.) air pollutants which would be created by this project at full buildout? Include short-term (construction) impacts: _____

APCD 44. Are there any sensitive receptors of air pollution located within one quarter mile of the project (e.g. schools, hospitals, etc.)? _____ Will the project generate any toxic/hazardous emissions? _____

APCD 45. What specific mobile/stationary source mitigation measures, if any, are proposed to reduce the air quality impact(s) of the project? Quantify any emission reductions and corresponding beneficial air quality impacts on a local/regional scale. _____

FIGURE A2.10   Sample environmental impact assessment questionnaire. (continued)

WATER

NOTE: *Based upon the type and complexity of the project, a detailed study of domestic water system capacity and/or groundwater impacts may be necessary).*

DPW 47. For what purpose is water presently used onsite? _____
_____
What and where is the existing source? _____
Is it treated water intended for domestic use? _____
What water sources will be used for this project? _____
Domestic: _____ Irrigation: _____
Fire Protection: _____ Other: _____
What is the projected peak water usage of the project? _____
Is the project within a public domestic water system district or service area? _____
If yes, will the public water supplier serve this project? _____
What is the proposed source of domestic water? _____
What is the projected peak water usage of the project? _____

DEH 48. Are there any wells on the site? _____ If so, describe depth, yield, contaminants, etc: _____
Show proposed well sites on the plan accompanying this application.

AESTHETICS

NOTE: *If the project has potential to visually impact an area's scenic quality, elevation drawings, photos or other depictions of the proposed project may be required.*

PLNG 49. Is the proposed project consistent/compatible with adjacent land uses and densities? _____
PLNG 50. Is the proposed project consistent/compatible with adjacent architectural styles? _____
PLNG 51. Would aesthetic features of the project (such as architecture, height, color, etc.) be subject to review? _____
By whom? _____
PLNG 52. Describe signs and lighting associated with the project: _____
PLNG 53. Is landscaping proposed? _____ If so, describe and indicate types and location of plants on a plan.

ARCHAEOLOGY/HISTORY

NOTE: *If the project site is on or near an historical or archaeological site, specific technical studies may be required for environmental determination.*

PLNG 54. What is the nearest historic site, state historic monument, national register district, or archaeological site?
_____
PLNG 55. How far away is it? _____
PLNG 56. Are there any historical, archaeological or culturally significant features on the site (i.e. old foundations, structures, Native American habitation sites, etc.)? _____

SEWAGE

NOTE: *Based upon the type and complexity of the project, a detailed analysis of sewage treatment and disposal alternatives may be necessary to make an environmental determination.*

DEH 57. How is sewage presently disposed of at the site? _____
DEH 58. How much wastewater is presently produced daily? _____
DEH 59. What is the proposed method of sewage disposal? _____
Is there a plan to protect groundwater from wastewater discharges? Yes____ No____ If yes, attach a draft of this plan.
DEH 60. How much wastewater would be produced daily? _____
DEH 61. List all unusual wastewater characteristics of the project, if any. What special treatment processes are necessary for these unusual wastes? _____

FIGURE A2.10 Sample environmental impact assessment questionnaire. (continued)

248 Appendix 2

                Will pre-treatment of wastewater be necessary? Yes____ No____ If yes, attach a description of pre-treatment processes and monitoring system.

DEH   62.   Is the groundwater level during the wettest time of the year less than 8 feet below the surface of the ground within the project area? _____

DEH   63.   Is this project located within a sewer district? _____
                If so, which district? _____ Can the district serve this project? ____

DEH   64.   Is there sewer in the area? _____

DEH   65.   What is the distance to the nearest sewer line? _____

HAZARDOUS MATERIALS

Hazardous materials are defined as any material that, because of its quantity, concentration, or physical or chemical characteristics, poses a significant present or potential hazard to human health and safety or to the environment if released into the workplace or the environment. "Hazardous materials" include, but are not limited to, hazardous substances, hazardous waste, and any material which a handler or the administering agency has a reasonable basis for believing that it would be injurious to the health and safety of persons or harmful to the environment if released into the workplace or the environment (including oils, lubricants, and fuels).

DEH   66.   Will the proposed project involve the handling, storage or transportation of hazardous materials? Yes____ No____

DEH   67.   If yes, will it involve the handling, storage, or transportation at any one time of more than 55 gallons, 500 pounds, or 200 cubic feet (at standard temperature and pressure) of a product or formulation containing hazardous materials? Yes____ No____

DEH   68.   If you answered yes to question #66, do you store any of these materials in underground storage tanks? Yes____ No____ If yes, please contact the Environmental Health Division at (916) 889-7335 for an explanation of additional requirements.

SOLID WASTE

DEH   69.   What types of solid waste will be produced? _____
                How much? _____ How will it be disposed of? _____

PARKS/RECREATION

PLNG   70.   How close is the project to the nearest public park or recreation area? _____
                Name the area _____

SOCIAL IMPACT

PLNG   71.   How many new residents will the project generate? _____

PLNG   72.   Will the project displace or require relocation of any residential units? _____

PLNG   73.   What changes in character of the neighborhood (surrounding uses such as pastures, farmland, residential) would the project cause? _____

PLNG   74.   Would the project create/destroy job opportunities? _____

PLNG   75.   Will the proposed development displace any currently productive use? _____
                If yes, describe: _____

TRANSPORTATION/CIRCULATION

Note: Detailed Traffic Studies prepared by a qualified consultant may be required following review of the information presented below.

DPW   76.   Does the proposed project front on a County road or State Highway? Yes____ No____
                If yes, what is the name of the road? _____

DPW   77.   If no, what is the distance to the nearest County road? _____
                Name of road? _____

FIGURE A2.10   Sample environmental impact assessment questionnaire. (continued)

DPW 78. Would any non-auto traffic result from the project (trucks, trains, etc.)? Yes_____ No_____
If yes, describe type and volume: _____

DPW 79. What road standards are proposed within the development? _____
Show typical street section(s) on the site plan.

DPW 80. Will new entrances onto County roads be constructed? Yes_____ No_____
If yes, show location on the site plan.

DPW 81. Describe any proposed improvements to County roads and/or State Highways:
_____
_____

DPW 82. How much additional traffic is the project expected to generate? (Indicate average daily traffic (ADT), peak hour volumes, identify peak hours. Use Institute of Transportation Engineers' (ITE) trip generation rates where project specific data is unavailable): _____

DPW 83. Would any form of transit be used for traffic to/from the project site? _____

DPW 84. What are the expected peak hours of traffic to be caused by the development (i.e., Churches: Sundays, 8:00 a.m. to 1:00 p.m.; Offices: Monday through Friday, 8:00 a.m. to 9:00 a.m., and 4:00 p.m. to 6:00 p.m.)? _____
_____

DPW 85. Will project traffic affect an existing traffic signal, major street intersection, or freeway interchange? Yes_____ No_____. If yes, explain: _____

DPW 86. What bikeway, pedestrian, equestrian, or transit facilities are proposed with the project? _____
_____

Name and title (if any) of person completing this Questionnaire: _____

Signature: _____ Date: _____

Title: _____ Telephone: _____

FIGURE A2.10 Sample environmental impact assessment questionnaire. (continued)

Complete the Environmental Impact Assessment Questionnaire and submit 20 copies of this form, one Initial Project Application, the current filing fee, and set of maps. Please submit 20 maps no larger than 8½"x11" (or **folded** to that size), including one reduced. For subdivision proposals, all information required by Section 19.125 of the Subdivision Ordinance for tentative map submittals, must be included in addition to the information listed below. Also provide an **aerial photo** of the site with a scale of 1" = 100' or same scale as the proposed tentative map.

1. Boundary lines and dimensions of parcel(s).
2. Existing and proposed structures and their gross floor area in square feet, parking areas with spaces delineated, distance between structures and distance from property lines.
3. The approximate area of the parcel (in square feet or acres).
4. Names, locations and widths of all existing traveled ways, including driveways, streets, and rights-of-way on, or adjacent to the property.
5. Approximate locations and widths of all proposed streets, rights-of-way, driveways, and/or parking areas.
6. Approximate location and dimensions of all proposed and existing easements, wells, leach lines, seepage pits, or other underground structures.
7. Approximate location and dimensions of all proposed easements for utilities and drainage.
8. Approximate location of all creeks, drainage channels, riparian areas, and a general indication of the slope of the land and all trees of significant size.
9. Accurately plot, label, and show exact location of the base and drip lines of all protected trees (native trees 6" dbh or greater, or multi-trunk trees 10" dbh or greater) within 50 feet of any development activity (i.e. proposed structures, driveways, cuts/fills, underground utilities, etc.) pursuant to Placer County Code, Chapter 36 (Tree Ordinance). Note: A tree survey prepared by an I.S.A. certified arborist may be required. Verify with the Planning Department prior to submittal of this application.
10. North arrow and approximate scale of drawing.
11. Vicinity map which shows the location of the subject property in relation to existing County roads and adjacent properties sufficient to identify the property in the field for someone unfamiliar with the area. The distance to the closest intersection of County roads should be shown to the nearest 1/10th of a mile.
12. Assessor's parcel number, section, township, and range.
13. Name(s) of property owner(s) and applicant, if any.
14. An indication of any adjacent lands in the same ownership.
15. **For areas in the Tahoe Basin only:** Existing impervious surface area (sq. ft.):_____; proposed _____. Impervious surface area allowed (sq. ft.) _____.

FOR INFORMATION REGARDING PROJECTS WITH EFFECTS THAT ARE NORMALLY SIGNIFICANT, REFER TO SECTION 31.450B OF THE PLACER COUNTY ENVIRONMENTAL REVIEW ORDINANCE. APPLICANTS ARE ENCOURAGED TO CONTACT THE STAFF PLANNER ASSIGNED TO THE PROJECT AT THE EARLIEST OPPORTUNITY TO DETERMINE POSSIBLE NEED AND SCOPE OF ADDITIONAL STUDIES.

FIGURE A2.10 Sample environmental impact assessment questionnaire.

USDA
Form RD 1940-20
(Rev. 6-99)

Position 3

REQUEST FOR ENVIRONMENTAL INFORMATION

FORM APPROVED
OMB No. 0575-0094

Name of Project

Location

Item 1a. Has a Federal, State, or Local Environmental Impact Statement or Analysis been prepared for this project?
☐ Yes ☐ No ☐ Copy attached as EXHIBIT I-A.
1b. If "No." provide the information requested in Instructions as EXHIBIT I.

Item 2. The State Historic Preservation Officer (SHPO) has been provided a detailed project description and has been requested to submit comments to the appropriate Rural Development Office. ☐ Yes ☐ No   Date description submitted to SHPO _____

Item 3. Are any of the following land uses or environmental resources either to be affected by the proposal or located within or adjacent to the project site(s)? *(Check appropriate box for every item of the following checklist.)*

	Yes	No	Unknown		Yes	No	Unknown
1. Industrial	☐	☐	☐	19. Dunes	☐	☐	☐
2. Commercial	☐	☐	☐	20. Estuary	☐	☐	☐
3. Residential	☐	☐	☐	21. Wetlands	☐	☐	☐
4. Agricultural	☐	☐	☐	22. Floodplain	☐	☐	☐
5. Grazing	☐	☐	☐	23. Wilderness *(designated or proposed under the Wilderness Act)*	☐	☐	☐
6. Mining, Quarrying	☐	☐	☐				
7. Forests	☐	☐	☐	24. Wild or Scenic River *(proposed or designated under the Wild and Scenic Rivers Act)*	☐	☐	☐
8. Recreational	☐	☐	☐				
9. Transportation	☐	☐	☐	25. Historical, Archeological Sites *(Listed on the National Register of Historic Places or which may be eligible for listing)*	☐	☐	☐
10. Parks	☐	☐	☐				
11. Hospital	☐	☐	☐	26. Critical Habitats *(endangered/threatened species)*	☐	☐	☐
12. Schools	☐	☐	☐				
13. Open spaces	☐	☐	☐	27. Wildlife	☐	☐	☐
14. Aquifer Recharge Area	☐	☐	☐	28. Air Quality	☐	☐	☐
15. Steep Slopes	☐	☐	☐	29. Solid Waste Management	☐	☐	☐
16. Wildlife Refuge	☐	☐	☐	30. Energy Supplies	☐	☐	☐
17. Shoreline	☐	☐	☐	31. Natural Landmark *(Listed on National Registry of Natural Landmarks)*	☐	☐	☐
18. Beaches	☐	☐	☐	32. Coastal Barrier Resources System	☐	☐	☐

Item 4. Are any facilities under your ownership, lease, or supervision to be utilized in the accomplishment of this project, either listed or under consideration for listing on the Environmental Protection Agency's List of Violating Facilities? ☐ Yes ☐ No

_____   Signed: _____
(Date)   *(Applicant)*

_____
(Title)

According to the Paperwork Reduction Act of 1995, an agency may not conduct or sponsor, and a person is not required to respond to, a collection of information unless it displays a valid OMB control number. The valid OMB control number for this information collections is 0575-0094. The time required to complete this information collection is estimated to average 15 minutes per response, including the time for reviewing instructions, searching existing data sources, gathering and maintaining the data needed, and completing and reviewing the collection of information.

FIGURE A2.11 Sample request for environmental information.

Plan Name:_____   Plan Number:_____

                                                Escrow Number:_____

CONSERVATION AGREEMENT

THIS AGREEMENT, made this _____ day of _____, 200_____ by and between _____ hereinafter called "Developer," party of the first part, and the Board of Supervisors of Prince William County, Virginia, hereinafter called "County," party of the second part:

WITNESSETH:

WHEREAS, Developer, desires approval of its site development plan, consisting of grading plans (which are part of the approved subdivision or site plans), erosion and sediment control plans, and/or landscaping plans, sewer, water and drainage plans (hereinafter collectively referred to as "plan"), for a project ("the project") known as _____. Said conservation plan also includes all provisions for conservation measures as required by the Code of Virginia, the Subdivision or Zoning Ordinance, and the Design and Construction Standards Manual; and

WHEREAS, the Developer intends to complete all of the development work contained on the approved subdivision or site plans, including but not limited to roads, sewer systems, water systems, storm water drainage systems, etc., at the project; and

WHEREAS, pursuant to Va. Code § 10.1-560 et seq. and other statutory authority, the County desires to ensure the proper installation, maintenance and adequate performance of such plan during the development process.

NOW, THEREFORE, for and in consideration of the foregoing premises and the following terms and conditions, and further in consideration of the approval of the aforesaid plan by the County and the issuance of a site preparation permit for the work proposed to be done thereunder, the parties hereto agree as follows:

1. Developer has provided guarantee in the amount of $_____ to the COUNTY in the form of one of the following which may be used for the purposes set forth in this Agreement:

    a) Cash deposit with Prince William County, receipt #_____. The sum deposited under this Agreement shall be placed in an interest-bearing account and the interest thereon shall accrue, up to a maximum of one year; or

    b) Letter of Credit # _____
       from (Name of Institution)_____.

2. In the event that measures for conservation as provided for in the plan referred to herein, or on any approved revision hereof, are not constructed or installed, the County shall give the Developer notice of violation and an opportunity to comply, and upon failure of the Developer to comply within the time period allowed by the County in its notice, the County shall have the right to enter upon Developer's property and shall construct such measures or do such other work as may be necessary, according to the plan to stabilize and make the site safe.

3. In the event the Director of the Department of Public Works or his designee determines that immediate construction or installation of conservation measures is required during the development process to prevent adverse sedimentation or erosion or to protect the public health, safety or welfare, the County shall give the Developer notice of such determination and an opportunity to construct or install conservation measures within a reasonable time period. Upon failure of the Developer to comply within the time period allowed by the County in its notice, the County shall enter upon Developer's property and construct such measures or do such work as may be necessary.

4. In the event the plan has been installed or constructed according to design, but fails, or inadequately effectuates the conservation measures required by County standards, or inadequately controls sediment or erosion; the Developer agrees to submit a revision to the plan and immediately institute measures to effectuate such measures or control upon notice of such event(s) by the County. In the event Developer fails to do so within the time period allowed by the County in its notice, the County may revise the plan and may enter upon Developer's property to construct the necessary measures.

5. In the event sedimentation and/or erosion from the property adversely affects downstream drainage, any adjacent property owner, or any street, road, highway or other public easement, the County may give the Developer notice of violation and an opportunity to comply, and upon failure of the Developer to comply within the time period allowed by the County in its notice, the County shall have the right to enter upon Developer's property to take such steps as may be necessary to prevent future off-site or on-site sedimentation or erosion, repair or clean up any off-site or on-site damage, or install any appropriate conservation measures.

6. The County shall give the Developer notice in the event tree protection or other conservation measures are not installed, damaged trees are not repaired, dead, dying or hazardous trees or branches within and contiguous to the development areas are not removed, or trees or other conservation measures required by the plan, or required revision, are not installed as specified on the plan, or required revision. If the Developer fails to comply within the time period allowed by the County in its notice, the County shall enter upon the Developer's property to perform such work.

7. In the event County performs work of any nature, including labor, use of equipment, and materials under the provisions of Paragraphs 2, 3, 4, 5 and 6 above, either by use of public forces or by private contract, it shall either (a) use the sum deposited herewith under Paragraph 1(a) and any accrued interest to pay for such work, or (b) draw on the letter of credit provided pursuant to Paragraph 1(b) to pay for such work. The Developer shall be sent notice when such sums are used.

8. In the event any portion of any guarantee provided hereunder is used by the County pursuant to this Agreement, Developer agrees to provide additional or replacement guarantee within ten (10) days of such use in an amount sufficient to restore the guarantee to an amount existing prior to the County's use of such guarantee.

9. It is expressly agreed by all parties hereto that it is the purpose and intent of this Agreement to conserve and protect the land, water, air and other natural resources of the Commonwealth and to ensure the proper construction, installation, maintenance, and performance of conservation measures provided by the plan or revisions thereof, and for the clean-up or repair of all damage on-site and off-site due to failed conservation measures, lack of conservation measures, or to erosion or sedimentation. This Agreement shall not be deemed to create or affect any liability of the County for any failure, lack of installation or damage alleged to result from or be caused by lack of conservation measures or by failed conservation measures, or by erosion or sedimentation.

10. The County shall hold the guarantee until it is satisfied that no further land-disturbing activity will be or is necessary to be taken on site in conjunction with the site preparation permit, all required conservation measures have been placed or installed and the County is satisfied that any required clean-up or repairs have been made. When these conditions are met, and in the event the guarantee is not used by the County as part of the cost of completion of development improvements (including required fees), or to restore the balance of

FIGURE A2.12   Sample conservation agreement. (continued)

any other conservation agreement deposit between this Developer and the County to its required level, all guarantee remaining after disbursement, if any, shall be released in writing by the County, through its agent, the Director of Planning.

11. All notices to be given with respect to this Agreement shall be in writing. Each notice shall be sent by registered or certified mail postage prepaid and return receipt requested, to the party to be notified at the address set forth herein or at such other address as either party may from time to time designate in writing, or by delivery at the site of the permitted activities to the agent or employee of the permittee supervising such activities. Every notice shall be deemed to have been given at the time it shall be deposited in the United States mails in the manner prescribed herein. Nothing contained herein shall be construed to preclude personal service of any notice in the manner prescribed for personal service of a summons or other legal process.

12. In the event Developer fails to comply with any provision of this Agreement and the County initiates legal proceedings to enforce its provisions, the County shall be entitled to receive all foreseeable damages, including, but not limited to, costs of engineering, design, construction, administration court costs and reasonable attorney's fees.

13. In conjunction with or subsequent to a notice to comply, the County may issue an order requiring that all or part of the land-disturbing activities permitted on the site be stopped until the specified corrective measures have been taken. Where the alleged noncompliance is causing or is in imminent danger of causing harmful erosion of lands or sediment deposition in waters or imperils the safety and welfare of the citizens of Prince William County within the Commonwealth, such an order may be issued without regard to whether the Developer has been issued a notice to comply. Otherwise, such an order may be issued only after the permittee has failed to comply with such a notice to comply. The order shall be served in the same manner as a notice to comply and shall remain in effect for a period of seven days from the date of service pending application by the County or the permit holder for appropriate relief to the Circuit Court. The order shall be lifted immediately following completion of the corrective action. Nothing in this paragraph shall prevent the County from taking any other action specified by law.

IN WITNESS of all which, the parties hereto have caused this Agreement to be executed on their behalf.

_____   _____
                                                                              Developer

_____   _____
Federal Tax I.D. or S.S.N.   Street Number

                                                                   _____
                                                                   City,   State,   Zip Code

_____   _____
Signature   Signature

_____   _____
Print Name and Title   Print Name, Title, and Phone Number

STATE OF VIRGINIA
COUNTY OF PRINCE WILLIAM

The above agreement was subscribed and confirmed by oath or affirmation before me this _____ day of _____,

200___ in the State of _____. My commission expires: _____.

                                                                   _____
                                                                                    NOTARY PUBLIC

BOARD OF SUPERVISORS OF PRINCE WILLIAM COUNTY

                                                    By: _____
                                                                              Bond Administrator

STATE OF VIRGINIA
COUNTY OF PRINCE WILLIAM

I, _____, a Notary Public in and for the State of Virginia, County of Prince William,

whose commission expires the _____ day of _____, 200 _____, acknowledges that

_____, Bond Administrator appeared before me this _____ day of

_____, 200___.

                                    Given under my hand this _____ day of _____, 200_____

                                                                   _____
                                                                                    NOTARY PUBLIC

Updated 3/3/03

FIGURE A2.12 Sample conservation agreement.

254 Appendix 2

Prince William County
Flood Hazard Use Permit Application

Property Address: _____ GPIN: _____-_____-_____

Subdivision: _____ Section and/or Phase: _____

Applicant: _____ Telephone Number: _____

Purpose of application: _____

_____

I, hereby certify that I have the authority to make the foregoing application, that the information submitted is correct, and that the construction will conform with the regulations in the Design and Construction Standards Manual, any applicable Federal, State, or local statutes, including the Building Code Ordinance, and the Virginia Contractor's Registration Law.

I further certify that the floodplain and/or floodway limits will be clearly demarcated in the vicinity of the construction area and that a copy of the approved permit application and one set of the approved plans shall be maintained at the site during construction.

I also understand that the issuance of an occupancy permit for any structure built within the flood hazard area shall be contingent upon the submission of a Federal Emergency Management Agency Elevation Certificate certified by a professional engineer or land surveyor.

_____   _____   _____
Applicant's Signature   Print Name   Date

☐   1. Application is approved as submitted.

☐   2. Application is approved with the following conditions: _____
     _____
     _____

☐   3. Application is denied.

_____   _____   _____
Authorized Signature   Title   Date

Planning Office Use:

File Number: _____ Permit Number: _____ Date: _____

FIGURE A2.13   Sample flood hazard use permit application.

DECLARATION OF LAND RESTRICTIONS FOR FLOOD MANAGEMENT

This DECLARATION made this _____ day of _____, 200__, by _____ Owners, and having an address at _____.

WITNESSETH:

WHEREAS, the Owner is the record owner of all that real property located on (street), (city), Election District __, designated in the 200__ Tax Records as Map __, Parcel ___, and being that same property acquired by the Owner by deed dated _____, 199_, and recorded among the Land Records of Allegany County, Maryland, at Libre ___ /Folio ____,

WHEREAS, the Owner has applied for a permit, conditioned permit, or variance to place a structure on that property that either (1) does not conform, or (2) may be made noncompliant by later conversion, to the strict elevation requirements outlined in the Floodplain Management Ordinance of Allegany County.

WHEREAS, the Owner agrees to record this DECLARATION and certifies and declares that the following covenants, conditions, and restrictions are placed on the affected property as a condition of granting the Permit, and affects rights and obligations of the Owner and shall be binding on the Owner, his heirs, personal representatives, successors, and assigns.

UPON THE TERMS AND SUBJECT TO THE CONDITIONS, as follows:

1. The structure or part thereof to which these conditions apply is/are: _____ _____, further described as floodplain management site plan file number ____FP_____, being a requirement of Land Use Permit Number _____.

2. This structure or part thereof has been allowed without strict conformance with the elevation requirement of the Ordinance. Conversion to habitable space shall not occur unless the enclosed area below the Flood Protection Elevation is brought into full compliance with this Ordinance. At this site, the Flood Protection Elevation is (undetermined)+/- feet above mean sea level, National Geodetic Vertical Datum.

3. Enclosed areas below the Flood Protection Elevation shall be used solely for parking of vehicles, limited storage, or access to the floor above. All interior walls, ceilings, and floors below the Flood Protection Elevation shall be unfinished and constructed of flood resistant materials. Mechanical, electrical, or plumbing devices shall not be installed below the Flood Protection Elevation.

4. The walls of the enclosed areas below the Flood Protection Elevation shall be equipped with at least two vents which permit the automatic entry and exit of flood waters with total openings of at least one square inch for every square foot of enclosed area below flood level. The vents shall be on two different walls, and the bottoms of the vents shall be no more than one foot above grade.

FIGURE A2.14 Sample declaration of land restrictions for flood management. (continued)

Declaration of Land Restrictions
MM/DD/YY
Page 2 of 2

5. Any alterations or changes from these conditions constitute a violation of the Permit and may render the structure uninsurable or increase the cost for flood insurance. The jurisdiction issuing the Permit and enforcing the Ordinance may to take any appropriate legal action to correct any violation.

6. Other conditions: n/a

      IN WITNESS THEREOF, this instrument is executed this _____th day of _____ 200__.

ATTEST:

                                                  _____ (Seal)
                                                  Owner

                                                  _____ (Seal)
                                                  Owner

NOTARY: STATE OF _____, County of _____, TO WIT:

      I hereby certify that on this _____ day of _____, 200 ____ , before me the subscriber, a Notary Public of the State aforesaid, personally appeared _____ and _____, known to me, or satisfactorily proven to be the person(s) whose name is subscribed to the foregoing instrument, who acknowledged that he has executed it for the purposes therein set forth, and that it is his act and deed.

      In witness whereof, I have set my hand and Notarial Seal, the day and year first written above.

My Commission expires on _____   _____
                                                          Notary Public

FIGURE A2.14 Sample declaration of land restrictions for flood management.

USDA	U.S. DEPARTMENT OF AGRICULTURE	Form Approved
Form RD 1924-25	RURAL DEVELOPMENT	OMB No. 0575-0042
(Rev. 7-99)	FARM SERVICE AGENCY	
	PLAN CERTIFICATION	

(Property Name/Applicants Name and Case Number)

(Property Address)	(City)
(County)	(State)

BUILDING TYPE: ☐ Single Family ☐ Multi-Family

PLANS: ☐ Original ☐ Modifications

I, _____ being a _____
     (type or print)   (licensed architect, engineer, or authorized building official, etc.)

in the State of _____ , hereby certify that I have reviewed:

☐ the plans and specifications dated _____ prepared by _____
                                                                                        (name of firm or individual)
for the above property

☐ the thermal performance plans, specifications and calculations dated _____
prepared by _____ for the above property
              (name of firm or individual)

☐ the seismic design (plans and specifications) dated _____ prepared by
_____ for the above property
(name of firm or individual)

☐ modifications listed below, that have been clearly indicated on the drawings and specifications
dated _____ prepared by _____ and certified by
                                                    (name of firm or individual)
_____ and related to the above property.
(name of firm or individual)

MODIFICATIONS _____

FIGURE A2.15 Sample plan certification form. (continued)

Based upon this review, to the best of my/our knowledge, information, and belief, these documents comply with the:

_____ and
(name and edition of the applicable development standard)

_____
(name and edition of the applicable energy standards/requirements
in accordance with RD Instruction 1924-A, Exhibit D)

designated as the applicable Rural Development or Farm Service Agency development standards for this project.

I understand the purpose of this certification is to induce United States Government to finance the construction of the above project and plan. I further understand that false certification constitutes a violation of 18 U.S.C. Section 1001 punishable by fine and/or imprisonment and, in addition, may result in debarment from participating in future government programs.

_____   _____
(Signature)   (Date)

_____   _____
(Type or print name)   (Professional Registration No.)

_____   _____
(Title)   (Expiration Date if applicable)

_____
(Area Code + Telephone Number)

FIGURE A2.15 Sample plan certification form.

Clearwater
CITY OF

Planning Department
100 South Myrtle Avenue
Clearwater, Florida 33756
Telephone: 727-562-4567
Fax: 727-562-4865

☐ SUBMIT ORIGINAL <u>SIGNED AND NOTARIZED</u> APPLICATION

☐ SUBMIT 12 COPIES OF THE ORIGINAL APPLICATION including folded site plans

☐ SUBMIT APPLICATION FEE $ _____

```
CASE #: _____
DATE RECEIVED: _____
RECEIVED BY (staff initials): _____
ATLAS PAGE #: _____
ZONING DISTRICT: _____
LAND USE CLASSIFICATION: _____
ZONING & LAND USE CLASSIFICATION OF
ADJACENT PROPERTIES:
    NORTH: _____  _____
    SOUTH: _____  _____
    WEST:  _____  _____
    EAST:  _____  _____
```

DEVELOPMENT AGREEMENT APPLICATION
(Revised 05/22/02)

~PLEASE TYPE OR PRINT~

A. APPLICANT, PROPERTY OWNER AND AGENT INFORMATION: (Section 4-202.A)

APPLICANT NAME: _____

MAILING ADDRESS: _____

PHONE NUMBER: _____   FAX NUMBER: _____

CELL NUMBER: _____   EMAIL ADDRESS: _____

PROPERTY OWNER(S): _____
(Must include ALL owners)

AGENT NAME: _____

MAILING ADDRESS: _____

PHONE NUMBER: _____   FAX NUMBER: _____

B. PROPOSED DEVELOPMENT INFORMATION:

STREET ADDRESS: _____

LEGAL DESCRIPTION: _____

PARCEL NUMBER: _____

PARCEL SIZE: _____
(acres, square feet)

PROPOSED USE AND SIZE: _____
(number of dwelling units, hotel rooms or square footage of nonresidential use)

DESCRIPTION OF ANY RELATED REQUEST(S): _____
(approval of a development include all requested code deviations; e.g. reduction in required number of parking spaces, specific use, etc.)

DOES THIS APPLICATION INVOLVE THE TRANSFER OF DEVELOPMENT RIGHTS (TDR), A PREVIOUSLY APPROVED PLANNED UNIT DEVELOPMENT, OR A PREVIOUSLY APPROVED (CERTIFIED) SITE PLAN? YES ____ NO ____ (if yes, attach a copy of the applicable documents)

FIGURE A2.16 Sample development agreement application. (continued)

B.2 DEVELOPMENT AGREEMENTS SUPPLEMENTAL SUBMITTAL REQUIREMENTS: (Section 4-606.B)

An application for approval of a development agreement shall be accompanied by the following (use separate sheets or include in a formal report):

- ☐ STATEMENT OF THE REQUESTED DURATION OF THE DEVELOPMENT AGREEMENT, WHICH SHALL NOT EXCEED TEN YEARS
- ☐ DESCRIPTION OF ALL EXISTING AND PROPOSED PUBLIC FACILITIES AND SERVICES THAT SERVE OR WILL SERVE THE DEVELOPMENT;
- ☐ DESCRIPTION OF THE USES DESIRED TO BE PERMITTED ON THE LAND, INCLUDING POPULATION DENSITIES AND BUILDING INTENSITIES AND HEIGHTS;
- ☐ INDENTIFICATION OF ZONING DISTRICT CHANGES, CODE AMENDMENTS THAT WILL BE REQUIRED IF THE PROPOSED DEVELOPMENT PROPOSAL WERE TO BE APPROVED;
- ☐ ZONING AND LAND USE CATEGORIES OF ALL ADJOINING PROPERTIES;
- ☐ COMPLETE NAMES AND ADDRESSES OF ALL OWNERS OR PROPERTIES ABUTTING OR LYING WITHIN 200 FEET OF THE SUBJECT PROPERTY AS CURRENTLY LISTED IN THE COUNTY RECORDS AS OF ONE WEEK PRIOR TO THE FILING OF AN APPLICATION.

C. PROOF OF OWNERSHIP: (Section 4-202.A)

- ☐ SUBMIT A COPY OF THE TITLE OR DEED TO THE PROPERTY OR PROVIDE OWNER SIGNATURE ON PAGE OF THIS APPLICATION

D. WRITTEN SUBMITTAL REQUIREMENTS: (Section 4-606.G)

- ☐ Provide the following contents to the development agreement, as follows:

Contents. The approved development agreement shall contain, at a minimum, the following information:

a. A legal description of the land subject to the development agreement.

b. The names of all persons having legal or equitable ownership of the land.

c. The duration of the development agreement, which shall not exceed ten years.

d. The development uses proposed for the land, including population densities, building intensities and building height.

e. A description of the public facilities and services that will serve the development, including who shall provide such public facilities and services; the date any new public facilities and services, if needed, will be constructed; who shall bear the expense of construction of any new public facilities and services; and a schedule to assure that the public facilities and services are available concurrent with the impacts of the development. The development agreement shall provide for a cashier's check, a payment and performance bond or letter of credit in the amount of 115 percent of the estimated cost of the public facilities and services, to be deposited with the city to secure construction of any new public facilities and services required to be constructed by the development agreement. The development agreement shall provide that such construction shall be completed prior to the issuance of any certificate of occupancy.

f. A description of any reservation or dedication of land for public purposes.

g. A description of all local development approvals approved or needed to be approved for the development.

h. A finding that the development approvals as proposed is consistent with the comprehensive plan and the community development code. Additionally, a finding that the requirements for concurrency as set forth in Article 4 Division 10 of these regulations have been satisfied.

i. A description of any conditions, terms, restrictions or other requirements determined to be necessary by the city commission for the public health, safety or welfare of the citizens of the City of Clearwater. Such conditions, terms, restrictions or other requirements may be supplemental to requirements in existing codes or ordinances of the city.

j. A statement indicating that the failure of the development agreement to address a particular permit, condition, term or restriction shall not relieve the developer of the necessity of complying with the law governing said permitting requirements, conditions, terms or restrictions.

k. The development agreement may provide, in the discretion of the City Commission, that the entire development or any phase thereof be commenced or be completed within a specific period of time. The development agreement may provide for liquidated damages, the denial of future development approvals, the termination of the development agreement, or the withholding of certificates of occupancy for the failure of the developer to comply with any such deadline.

l. A statement that the burdens of the development agreement shall be binding upon, and the benefits of the development agreement shall inure to, all successors in interest to the parties to the development agreement.

m. All development agreements shall specifically state that subsequently adopted ordinances and codes of the city which are of general application not governing the development of land shall be applicable to the lands subject to the development agreement, and that such modifications are specifically anticipated in the development agreement.

FIGURE A2.16 Sample development agreement application. (continued)

E. SUPPLEMENTAL SUBMITTAL REQUIREMENTS: (Section 4-202.A)

- ☐ SIGNED AND SEALED SURVEY (including legal description of property) – **One original and 12 copies;**
- ☐ COPY OF RECORDED PLAT, as applicable;
- ☐ PRELIMINARY PLAT, as required;
- ☐ LOCATION MAP OF THE PROPERTY.
- ☐ TREE SURVEY (including existing trees on site and within 25' of the adjacent site, by species, size (DBH 4" or greater), and location, including drip lines.)
- ☐ GRADING PLAN, as applicable;

F. SITE PLAN SUBMITTAL REQUIREMENTS: (Section 4-202.A)

- ☐ SITE PLAN with the following information (not to exceed 24" x 36"):
  - ___ All dimensions;
  - ___ North arrow;
  - ___ Engineering bar scale (minimum scale one inch equals 50 feet), and date prepared;
  - ___ Location map;
  - ___ Index sheet referencing individual sheets included in package;
  - ___ Footprint and size of all buildings and structures;
  - ___ All required setbacks;
  - ___ All existing and proposed points of access;
  - ___ All required sight triangles;
  - ___ Identification of environmentally unique areas, such as watercourses, wetlands, tree masses, and specimen trees, including description and location of understory, ground cover vegetation and wildlife habitats, etc;
  - ___ Location of all public and private easements;
  - ___ Location of all street rights-of-way within and adjacent to the site;
  - ___ Location of existing public and private utilities, including fire hydrants, storm and sanitary sewer lines, manholes and lift stations, gas and water lines;
  - ___ All parking spaces, driveways, loading areas and vehicular use areas;
  - ___ Depiction by shading or crosshatching of all required parking lot interior landscaped areas;
  - ___ Location of all refuse collection facilities and all required screening (min. 10'x12' clear space);
  - ___ Location of all landscape material;
  - ___ Location of all onsite and offsite storm-water management facilities;
  - ___ Location of all outdoor lighting fixtures; and
  - ___ Location of all existing and proposed sidewalks.
- ☐ SITE DATA TABLE for existing, required, and proposed development, in written/tabular form:
  - ___ Land area in square feet and acres;
  - ___ Number of dwelling units proposed;
  - ___ Gross floor area devoted to each use;
  - ___ Parking spaces: total number, presented in tabular form with the number of required spaces;
  - ___ Total paved area, including all paved parking spaces and driveways, expressed in square feet and percentage of the paved vehicular area;
  - ___ Size and species of all landscape material;
  - ___ Official records book and page numbers of all existing utility easement;
  - ___ Building and structure heights
  - ___ Impermeable surface ratio (I.S.R.); and
  - ___ Floor area ratio (F.A.R.) for all nonresidential uses.
- ☐ REDUCED SITE PLAN to scale (8 ½ X 11) and color rendering if possible;
- ☐ FOR DEVELOPMENTS OVER ONE ACRE, provide the following additional information on site plan:
  - ___ One-foot contours or spot elevations on site;
  - ___ Offsite elevations if required to evaluate the proposed stormwater management for the parcel;
  - ___ All open space areas;
  - ___ Location of all earth or water retaining walls and earth berms;
  - ___ Lot lines and building lines (dimensioned);
  - ___ Streets and drives (dimensioned);
  - ___ Building and structural setbacks (dimensioned);
  - ___ Structural overhangs;
  - ___ Tree Inventory; prepared by a "certified arborist", of all trees 8" DBH or greater, reflecting size, canopy (drip lines) and condition of such trees.

FIGURE A2.16 Sample development agreement application. (continued)

G. LANDSCAPING PLAN SUBMITTAL REQUIREMENTS: (Section 4-1102.A)

- ☐ LANDSCAPE PLAN:
  - ___ All existing and proposed structures;
  - ___ Names of abutting streets;
  - ___ Drainage and retention areas including swales, side slopes and bottom elevations;
  - ___ Delineation and dimensions of all required perimeter landscape buffers;
  - ___ Sight visibility triangles;
  - ___ Delineation and dimensions of all parking areas including landscaping islands and curbing;
  - ___ Proposed and required parking spaces;
  - ___ Existing trees on-site and immediately adjacent to the site, by species, size and locations, including dripline;
  - ___ Location, size, description, specifications and quantities of all existing and proposed landscape materials, including botanical and common names;
  - ___ Typical planting details for trees, palms, shrubs and ground cover plants including instructions, soil mixes, backfilling, mulching and protective measures;
  - ___ Interior landscaping areas hatched and/or shaded and labeled and interior landscape coverage, expressing in both square feet and percentage covered;
  - ___ Conditions of a previous development approval (e.g. conditions imposed by the Community Development Board);
  - ___ Irrigation notes.
- ☐ REDUCED LANDSCAPE PLAN to scale (8 ½ X 11) (color rendering if possible);
- ☐ IRRIGATION PLAN (required for level two and three approval);
- ☐ COMPREHENSIVE LANDSCAPE PROGRAM application, as applicable.

H. BUILDING ELEVATION PLAN SUBMITTAL REQUIREMENTS: (Section 4-202.A.23)

Required in the event the application includes a development where design standards are in issue (e.g. Tourist and Downtown Districts) or as part of a Comprehensive Infill Redevelopment Project or a Residential Infill Project.

- ☐ BUILDING ELEVATION DRAWINGS – all sides of all buildings including height dimensions, colors and materials;
- ☐ REDUCED BUILDING ELEVATIONS – four sides of building with colors and materials to scale (8 ½ X 11) (black and white and color rendering, if possible) as required.

I. SIGNAGE: (Division 19. SIGNS / Section 3-1806)

- ☐ Comprehensive Sign Program application, as applicable (separate application and fee required).
- ☐ Reduced signage proposal (8 ½ X 11) (color), if submitting Comprehensive Sign Program application.

J. TRAFFIC IMPACT STUDY: (Section 4-801.C)

- ☐ Include as required if proposed development will degrade the acceptable level of service for any roadway as adopted in the Comprehensive Plan. Trip generation shall be based on the most recent edition of the Institute of Transportation Engineer's Trip General Manual. Refer to Section 4-801 C of the Community Development Code for exceptions to this requirement.

K. SIGNATURE:

I, the undersigned, acknowledge that all representations made in this application are true and accurate to the best of my knowledge and authorize City representatives to visit and photograph the property described in this application.

STATE OF FLORIDA, COUNTY OF PINELLAS
Sworn to and subscribed before me this _____ day of _____, A.D. 20___ to me and/or by _____, who is personally known has produced _____ as identification.

_____
Signature of property owner or representative

Notary public,
My commission expires:

FIGURE A2.16 Sample development agreement application. (continued)

L. AFFIDAVIT TO AUTHORIZE AGENT:

_____
(Names of all property owners)

1. That (I am/we are) the owner(s) and record title holder(s) of the following described property (address or general location):

   _____
   _____
   _____

2. That this property constitutes the property for which a request for a: (describe request)

   _____
   _____

3. That the undersigned (has/have) appointed and (does/do) appoint: _____

   as (his/their) agent(s) to execute any petitions or other documents necessary to affect such petition;

4. That this affidavit has been executed to induce the City of Clearwater, Florida to consider and act on the above described property;

5. That site visits to the property are necessary by City representatives in order to process this application and the owner authorizes City representatives to visit and photograph the property described in this application;

6. That (I/we), the undersigned authority, hereby certify that the foregoing is true and correct.

                                                            _____
                                                            Property Owner
                                                            _____
                                                            Property Owner

STATE OF FLORIDA,
COUNTY OF PINELLAS

Before me the undersigned, an officer duly commissioned by the laws of the State of Florida, on this _____ day of _____, _____ personally appeared _____ who having been first duly sworn Deposes and says that he/she fully understands the contents of the affidavit that he/she signed.

                                                            _____
My Commission Expires: Notary Public

FIGURE A2.16 Sample development agreement application.

264 Appendix 2

CITY OF CLEARWATER
APPLICATION FOR PLAT APPROVAL
PLANNING & DEVELOPMENT SERVICES ADMINISTRATION
MUNICIPAL SERVICES BUILDING, 100 SOUTH MYRTLE AVENUE, 2nd FLOOR
PHONE (727)-562-4567   FAX (727) 562-4576

PROPERTY OWNER'S NAME : _____

ADDRESS : _____

PHONE NUMBER : _____ FAX NUMBER : _____

APPLICANT'S NAME : _____

ADDRESS : _____

PHONE NUMBER : _____ FAX NUMBER : _____

AGENT NAME : _____

ADDRESS : _____

PHONE NUMBER : _____ FAX NUMBER : _____

I, the undersigned, acknowledge that all representations made in this application are true and accurate to the best of my knowledge.
   by _____, who is personally known to

_____
Signature of owner or representative

STATE OF FLORIDA, COUNTY OF PINELLAS
Sworn to and subscribed before me this ___ day of _____, A.D., 19___ to me and/or

me or has produced_____
as identification.

_____
Notary Public
my commission expires:

Ten (10) copies of the preliminary plat must be submitted.

The preliminary plat shall be prepared by a surveyor, architect, landscape architect or engineer drawn to a scale not smaller than 1: 100 and shall not exceed 24" X 36" and include the following information:

NORTH ARROW, SCALE AND DATE;

TITLE UNDER WHICH THE PROPOSED PLAT IS TO BE RECORDED;

NAME, ADDRESS AND TELEPHONE NUMBER OF THE PERSON PREPARING THE PLAT;

IDENTIFICATION CLEARLY STATING THAT THE DRAWING IS A PRELIMINARY PLAT;

LEGAL DESCRIPTION OF THE PROPERTY WITH U.S. SURVEY SECTION, TOWNSHIP AND RANGE LINES;

EXISTING AND PROPOSED RIGHTS-OF-WAY AND EASEMENTS;

PROPOSED STREET NAMES;

NAMES, APPROPRIATELY POSITIONED, OF ADJOINING PLATS;

DIMENSIONS AND AREA OF THE FOLLOWING:

    THE OVERALL PLAT AND EACH LOT
    STREETS
    RIGHTS-OF-WAY, INCLUDING RADII OF CUL-DE-SACS;
    COMMON OPEN SPACE OR OTHER LAND TO BE DEDICATED FOR A PUBLIC PURPOSE IF ANY.

FIGURE A2.17 Sample application for plat approval. (continued)

CITY OF CLEARWATER
PLANNING & DEVELOPMENT SERVICES
MUNICIPAL SERVICES BUILDING, 100 SOUTH MYRTLE AVENUE, 2nd FLOOR
PHONE (727)-562-4567   FAX (727) 562-4576
INFORMATION REQUIRED FOR SUBMITTAL OF
FINAL PLAT INSTRUCTIONS

The final plat shall be suitable for recording at the office of the clerk of the circuit court. It shall be prepared and sealed buy a land surveyor registered by the state and shall conform with the requirements of Florida Statute, Chapter 177, and the requirements of this subsection. It shall be drawn at a scale of one inch equals 50 feet or other scale determined appropriate by the city engineer. The overall sheet size of the plat shall be consistent with the standards established by the clerk of the circuit court for recording. Each sheet shall be provided with a one-inch margin on each of three sides and a three-inch margin on the left side of the plat for binding purposes.

Eighteen (18) copies of the Final Plat must be submitted.

ALL FINAL PLATS MUST CONTAIN THE FOLLOWING INFORMATION:

NAME OF PLAT;

LOCATION OF THE PLAT BY U.S. SURVEY SYSTEM AND POLITICAL SUBDIVISION, INCLUDING   SECTION, TOWNSHIP, RANGE, COUNTY AND STATE;

NAMES OF EXISTING STREETS ABUTTING RO GIVING ACCESS TO THE PROPOSED PLAT;

ALL PLAT BOUNDARIES BASED ON AN ACCURATE TRANSVERSE, WITH ALL ANGULAR AND LINEAR DIMENSIONS SHOWN. ERROR OF ENCLOSURE OF SUCH BOUNDARY SURVEY SHALL NOT EXCEED ONE FOOT FOR EACH 10,000 FEET OF PERIMETER SURVEY;

ALL BLOCKS, LOTS, STREETS, CROSSWALKS, EASEMENTS AND WATERWAYS, WITHIN AND ADJACENT TO THE PLAT, ALL OF WHICH SHALL HAVE ALL ANGULAR AND LINEAR DIMENSIONS GIVEN AND   ALL   RADII, INTERNAL ANGLES, BEARINGS, POINTS OF CURVATURE, TANGENTS AND LENGTHS OF ALL CURVES, SO THAT NO DIMENSIONS OR DATA ARE MISSING WHICH ARE REQUIRED FOR THE   FUTURE LOCATION OF ANY OF THE CORNERS OR BOUNDARIES OF BLOCKS, LOTS OR STREETS, AS LISTED ABOVE. WHEN ANY LOT OR PORTION OF THE PLAT IS BOUNDED BY AN IRREGULAR LINE, THE MAJOR PORTION OF THAT LOT OR PLAT SHALL BE ENCLOSED BY A WITNESS LINE SHOWING COMPLETE DATA, WITH DISTANCES ALONG SUCH LINES EXTENDED BEYOND THE ENCLOSURE TO THE IRREGULAR BOUNDARY SHOWN WITH AS MUCH CERTAINTY AS CAN BE DETERMINED OR AS "MORE OR LESS", IF VARIABLE. ALL DIMENSIONS SHALL BE GIVEN TO THE NEAREST HUNDREDTH OF   A FOOT. TRUE ANGLES AND DISTANCES SHALL BE DRAWN TO THE NEAREST ESTABLISHED OFFICIAL MONUMENTS, NOT LESS THAN THREE OF WHICH SHALL BE ACCURATELY DESCRIBED ON THE PLAT. THE INTENDED USE OF ALL EASEMENTS SHALL BE CLEARLY STATED.

CURVILINEAR LOTS SHALL SHOW ARC DISTANCES, AND RADII, CHORD, AND CHORD BEARING. RADIAL LINES SHALL BE SO DESIGNATED. DIRECTION OF NONRADIAL LINES SHALL BE INDICATED;

SUFFICIENT ANGLES AND BEARINGS SHALL IDENTIFY THE DIRECTION OF ALL LINES AND SHALL BE SHOWN TO THE NEAREST SECOND;

FIGURE A2.17   Sample application for plat approval. (continued)

ALL RIGHT-OF-WAY CENTERLINES SHALL BE SHOWN WITH DISTANCES, ANGLES, BEARINGS OR AZIMUTH, POINTS OF CURVATURE, POINTS OF TANGENCY, POINTS OF REVERSE CURVATURE, POINTS OF COMPOUND CURVATURE, ARC DISTANCE, CENTRAL ANGLES, TANGENTS, RADII, CHORD, AND CHORD BEARING OR AZIMUTH, OR BOTH

ALL EASEMENTS OR RIGHTS-OF-WAY PROVIDED FOR PUBLIC SERVICES OR UTILITIES, AND ANY LIMITATIONS OF SUCH EASEMENTS;

ALL LOT NUMBERS AND LINES. LOT LINES SHALL BE MARKED WITH ACCURATE DIMENSIONS IN FEET AND HUNDREDTHS OF FEET, AND BEARINGS OR ANGLES TO STREET LINES;

ACCURATE DESCRIPTIONS OF ANY AREA TO BE DEDICATED OR RESERVED FOR PUBLIC USE WITH THE PURPOSE INDICATED THEREON;

TITLE, DATE OF SURVEY, GRAPHIC SCALE OF MAP AND NORTH ARROW. THE BEARING OR AZIMUTH REFERENCE SHALL BE CLEARLY STATED ON THE FACE OF THE PLAT IN THE NOTES OR LEGEND;

PERMANENT REFERENCE MONUMENTS SHALL BE PLACED IN ACCORDANCE WITH REQUIREMENTS OF THE SATE OF FLORIDA;

EACH PLAT SHALL SHOW A DESCRIPTION OF THE LANDS PLATTED, AND THE DESCRIPTION SHALL BE THE SAME IN THE TITLE CERTIFICATION. THE DESCRIPTION SHALL BE SO COMPLETE THAT FROM IT, WITHOUT REFERENCE TO THE PLAT, THE STARTING POINT AND BOUNDARY CAN BE DETERMINED;

THE CIRCUIT COURT CLERK'S CERTIFICATE AND THE LAND SURVEYOR'S CERTIFICATE AND SEAL; ALL SECTION LINES AND QUARTER SECTION LINES OCCURRING IN THE MAP OR PLAT SHALL BE INDICATED BY LINES DRAWN UPON THE MAP OR PLAT, WITH APPROPRIATE WORDS AND FIGURES. IF THE DESCRIPTION IS BY METES AND BOUNDS, THE POINT OF BEGINNING SHALL BE INDICATED, TOGETHER WITH ALL BEARINGS AND DISTANCES OF THE BOUNDARY LINES. IF THE PLATTED LANDS ARE IN A LAND GRANT OR ARE NOT INCLUDED IN THE SUBDIVISION OF GOVERNMENT SURVEYS, THEN THE BOUNDARIES ARE TO BE DEFINED BY METES AND BOUNDS AND COURSES. THE POINT OF BEGINNING IN THE DESCRIPTION SHALL BE TIED TO THE NEAREST GOVERNMENT CORNER OF OTHER RECORDED AND WELL-ESTABLISHED CORNER;

ALL CONTIGUOUS PROPERTIES SHALL BE IDENTIFIED BY PLAT TITLE, PLAT BOOK AND PAGE OR, IF UNPLATTED, LAND SHALL BE SO DESIGNED. IF THE AREA PLATTED IS A REPLATTING OF A PART OR THE WHOLE OF A PREVIOUSLY RECORDED PLAT, SUFFICIENT TIES SHALL BE SHOWN TO CONTROLLING LINES APPEARING ON THE EARLIER PLAT TO PERMIT AN OVERLAY TO BE MADE AND REFERENCE TO THE REPLATTING SHALL BE STATED AS A SUBTITLE FOLLOWING THE NAME OF THE PLAT WHEREVER IT APPEARS ON THE PLAT;

ALL LOTS SHALL BE NUMBERED EITHER BY PROGRESSIVE NUMBERS OR, IF IN BLOCKS, PROGRESSIVELY NUMBERED OR LETTERED IN EACH BLOCK, EXCEPT THAT BLOCKS IN NUMBER ADDITIONS BEARING THE SAME NAME MAY BE NUMBERED CONSECUTIVELY THROUGHOUT THE SEVERAL ADDITIONS;

PARK, RECREATION AND OPEN SPACE PARCELS SHALL BE SO DESIGNATED;

ALL INTERIOR EXCEPTED PARCELS SHALL BE CLEARLY INDICATED AND LABELED "NOT A PART OF THIS PLAT";

THE PURPOSE OF ALL AREAS DEDICATED MUST BE CLEARLY INDICATED OR STATED ON THE PLAT;

WHEN IT IS NOT POSSIBLE TO SHOW CURVE DETAIL INFORMATION ON THE MAP, A TABULAR FORM MAY BE USED.

FIGURE A2.17 Sample application for plat approval. (continued)

THE FOLLOWING DOCUMENTATION MUST BE SUBMITTED WITH THE FINAL PLAT:

A TITLE OPINION OF AN ATTORNEY LICENSED IN THE SATE OR A CERTIFICATION BY AN ABSTRACTOR OR A TITLE COMPANY STATING THAT THE COURT RECORDS IDENTIFY THAT THE TITLE TO THE LAND AS DESCRIBED AND SHOWN ON THE PLAT IS IN THE NAME OF THE PERSON EXECUTING THE DEDICATION. IN ADDITION, A DOCUMENT ENTITLED CONSENT TO PLATTING OF LANDS AND PARTIAL RELEASE OF MORTGAGE SHALL BE FILED TOGETHER WITH THE FINAL PLAT FOR EACH PERSON OR CORPORATION

HOLDING A MORTGAGE ON ALL LAND INCLUDED ON THE PLAT, WHERE SUCH PERSON HAS NOT SIGNED THE FINAL PLAT. THE TITLE OPINION OR CERTIFICATION SHALL SHOW ALL MORTGAGES NOT SATISFIED OR

RELEASED OF RECORD NOR OTHER WISE TERMINATED BY LAW;

CERTIFICATION BY A REGISTERED LAND SURVEYOR THAT THE PLAT REPRESENTS A SURVEY MADE BY THAT INDIVIDUAL, THAT ALL THE NECESSARY SURVEY MONUMENTS, LOT SIZES AND LOT DIMENSIONS ARE CORRECTLY SHOWN THEREON, AND THAT THE PLAT COMPLIES WITH ALL OF THE SURVEY REQUIREMENTS OF CHAPTER 177 AND THIS DEVELOPMENT CODE. IMPRESSED ON THE PLAT AND AFFIXED THERETO SHALL BE THE PERSONAL SEAL AND SIGNATURE TO THE REGISTERED LAND SURVEYOR INCLUDING THE REGISTRATION NUMBER OF THE SURVEYOR, BY WHOM OR UNDER WHOSE AUTHORITY AND DIRECTION THE PLAT WAS PREPARED;

A BOUNDARY SURVEY OF THE PLATTED LANDS. HOWEVER, A NEW BOUNDARY SURVEY FOR A REPLAT IS REQUIRED ONLY WHEN THE REPLAT AFFECTS ANY BOUNDARY OF THE PREVIOUSLY PLATTED PROPERTY OR WHEN IMPROVEMENTS HAVE BEEN MADE ON THE LANDS TO BE REPLATTED OR ADJOINING LANDS. THE BOUNDARY SURVEY MUST BE PERFORMED AND PREPARED UNDER THE RESPONSIBLE DIRECTION AND SUPERVISION OF A PROFESSIONAL SURVEYOR AND MAPPER PRECEDING THE INITIAL SUBMITTAL OF THE PLAT TO THE LOCAL GOVERNING BODY. THIS SUBSECTION DOES NOT RESTRICT A LEGAL ENTITY FROM EMPLOYING ONE PROFESSIONAL SURVEYOR AND MAPPER TO PERFORM AND PREPARE THE BOUNDARY SURVEY AND ANOTHER PROFESSIONAL SURVEYOR AND MAPPER TO PREPARE THE PLAT, EXCEPT THAT BOTH THE BOUNDARY SURVEY AND THE PLAT MUST BE UNDER THE SAME LEGAL ENTITY;

CERTIFICATION THAT ALL REAL ESTATE TAXES HAVE BEEN PAID;

EVERY PLAT OF A SUBDIVISION OR CONDOMINIUM FILED FOR RECORD SHALL INCLUDE ANY REQUIRED DEDICATION BY THE APPLICANT. THE DEDICATION SHALL BE EXECUTED BY ALL OWNERS HAVING A RECORD INTEREST IN THE LAND BEING PLATTED, IN THE SAME MANNER IN WHICH DEEDS ARE REQUIRED TO BE EXECUTED. ALL MORTGAGEES HAVING A RECORD INTEREST IN THE LAND PLATTED SHALL EXECUTE, IN THE SAME MANNER IN WHICH DEEDS ARE REQUIRED TO BE EXECUTED, EITHER THE DEDICATION CONTAINED ON THE PLAT OR IN A SEPARATE INSTRUMENT JOINING THE RATIFICATION OF THE PLAT AND ALL DEDICATION AND RESERVATIONS THEREON IN THE FORM OF A CONSENT TO PLAT FROM ALL MORTGAGE INTERESTS ACCEPTABLE TO THE CITY ATTORNEY. WHEN A TRACT OR PARCEL OF LAND HAS BEEN PLATTED AND A PLAT THEREOF BEARING THE DEDICATION EXECUTED BY THE DEVELOPER AND APPROVAL OF THE CITY HAS BEEN SECURED AND RECORDED IN COMPLIANCE WITH THIS DIVISION, ALL STREETS, ALLEYS, EASEMENTS, RIGHTS-OF-WAY AND PUBLIC AREAS SHOWN ON SUCH PLAT, UNLESS OTHERWISE STATED, SHALL BE DETERMINED TO HAVE BEEN DEDICATED TO THE PUBLIC FOR THE USES AND PURPOSES STATED THEREON, NOTWITHSTANDING ANY SEPARATE ACTION BY RESOLUTION OF THE CITY COMMISSION TO FORMALLY ACCEPT SUCH OFFERS OF DEDICATION;

ANY EXISTING OR PROPOSED PRIVATE RESTRICTION AND TRUSTEESHIPS AND THEIR PERIODS OF EXISTENCE SHALL BE FILED AS A SEPARATE INSTRUMENT AND REFERENCE TO SUCH INSTRUMENT SHALL BE NOTED ON THE FINAL PLAT;

FIGURE A2.17 Sample application for plat approval. (continued)

AFTER A FINAL PLAT HAS BEEN APPROVED, THREE PRINTS OF AS-BUILT DRAWINGS SHOWING THE IMPROVEMENTS THAT HAVE BEE CONSTRUCTED ACCORDING TO THE APPROVED SUBDIVISION CONSTRUCTION PLANS AND A COPY OF THE FINANCIAL GUARANTEE FOR COMPLETION OF REQUIRED IMPROVEMENTS SHALL BE FILED WITH THE CITY ENGINEER BEFORE SUCH PLAT SHALL BE RECORDED.

FINANCIAL GUARANTEE: UNLESS ALL REQUIRED IMPROVEMENTS HAVE BEEN SATISFACTORILY COMPLETED, AN ACCEPTABLE FINANCIAL GUARANTEE FOR REQUIRED IMPROVEMENTS SHALL ACCOMPANY EVERY PLAT WHICH IS TO BE RECORDED TO ENSURE THE ACTUAL SATISFACTORY COMPLETION OF CONSTRUCTION OF ALL REQUIRED IMPROVEMENTS WITHIN NOT MORE THAN TWO YEARS FOLLOWING THE DATE OF THE RECORDING, OR ONE YEAR IF SIDEWALKS ARE THE ONLY REQUIRED IMPROVEMENT TO BE COMPLETED FOLLOWING THE DATE OF RECORDING. AN ACCEPTABLE FINANCIAL GUARANTEE FOR REQUIRED IMPROVEMENTS SHALL BE IN AN AMOUNT NOT LESS THAN THE ESTIMATED COST OF THE IMPROVEMENTS, AS APPROVED BY THE CITY ENGINEER, AND MAY BE REQUIRED TO BE INCREASED IF THE CITY ENGINEER DETERMINES IT APPROPRIATE AND MAY BE REDUCED FROM TIME TO TIME IN PROPORTION TO THE WORK COMPLETED, AND MAY TAKE ONE OF THE FOLLOWING FORMS, SUBJECT TO THE APPROVAL OF THE CITY ENGINEER AND THE CITY ATTORNEY;

CASH, TO BE HELD IN A SEPARATE ESCROW ACCOUNT BY THE CITY; OR

AN IRREVOCABLE LETTER OF CREDIT WRITTEN BY A BANK CHARTERED BY THE SATE, THE UNITED STATES GOVERNMENT, OR ANY OTHER STATE OF THE UNITED STATES IF THE BANK IS AUTHORIZED TO DO BUSINESS IN THE STATE OF FLORIDA, AND ACCEPTABLE TO THE CITY MANAGER. THE LETTER OF CREDIT SHALL INCLUDE AMONG OTHER THINGS, AN EXPIRATION DATE NOT EARLIER THAN ONE YEAR FROM THE DATE OF ISSUANCE; A PROVISION REQUIRING THE ISSUER OF THE LETTER OF CREDIT TO GIVE AT LEAST 30 DAYS WRITTEN NOTICE TO THE CITY PRIOR TO EXPIRATION OR RENEWAL OF THE LETTER; AND A PROVISION THAT THE LETTER IS AUTOMATICALLY RENEWED FOR A PERIOD OF TIME EQUALING ITS ORIGINAL TERM OF THE REQUIRED NOTICE IS NOT GIVEN; OR

A SURETY BOND ISSUED BY A SURETY COMPANY AUTHORIZED TO DO BUSINESS IN THE STATE. THE SURETY BOND SHALL INCLUDE, AS A MINIMUM, THE PROVISION REQUIRED FOR LETTERS OF CREDIT.

FIGURE A2.17 Sample application for plat approval.

Clearwater
CITY OF

Planning Department
100 South Myrtle Avenue
Clearwater, Florida 33756
Telephone: 727-562-4567
Fax: 727-562-4865

CASE #:
DATE RECEIVED:
RECEIVED BY (staff initials):
ATLAS PAGE #:
ZONING DISTRICT:
LAND USE CLASSIFICATION:
SURROUNDING USES OF ADJACENT PROPERTIES:
NORTH:
SOUTH:
WEST:
EAST:

❑ SUBMIT ORIGINAL <u>SIGNED AND NOTARIZED</u> APPLICATION

❑ SUBMIT 12 COPIES OF THE ORIGINAL APPLICATION including
 1) collated, 2) stapled and 3) folded sets of site plans

❑ SUBMIT APPLICATION FEE $ _____

* NOTE: 13 TOTAL SETS OF INFORMATION REQUIRED (APPLICATIONS PLUS SITE PLANS SETS)

FLEXIBLE DEVELOPMENT APPLICATION
Comprehensive Infill Redevelopment Project (Revised 11/05/02)

~PLEASE TYPE OR PRINT~ use additional sheets as necessary

A. APPLICANT, PROPERTY OWNER AND AGENT INFORMATION: (Section 4-202.A)

APPLICANT NAME: _____

MAILING ADDRESS: _____

E-MAIL ADDRESS: _____ PHONE NUMBER: _____

CELL NUMBER: _____ FAX NUMBER: _____

PROPERTY OWNER(S): _____
 (Must include ALL owners)

AGENT NAME(S): _____

MAILING ADDRESS: _____

E-MAIL ADDRESS: _____ PHONE NUMBER: _____

CELL NUMBER: _____ FAX NUMBER: _____

B. PROPOSED DEVELOPMENT INFORMATION:

STREET ADDRESS of subject site: _____

LEGAL DESCRIPTION: _____
 (if not listed here, please note the location of this document in the submittal)

PARCEL NUMBER: _____

PARCEL SIZE: _____
 (acres, square feet)

PROPOSED USE AND SIZE: _____
 (number of dwelling units, hotel rooms or square footage of nonresidential use)

DESCRIPTION OF REQUEST(S): _____
 (include all requested code deviations; e.g. reduction in required number of parking spaces, specific use, etc.)

FIGURE A2.18 Sample flexible development application. (continued)

270 Appendix 2

DOES THIS APPLICATION INVOLVE THE TRANSFER OF DEVELOPMENT RIGHTS (TDR), A PREVIOUSLY APPROVED PLANNED UNIT DEVELOPMENT, OR A PREVIOUSLY APPROVED (CERTIFIED) SITE PLAN? YES ____ NO ____ (if yes, attach a copy of the applicable documents)

C. PROOF OF OWNERSHIP: (Section 4-202.A)

❏ SUBMIT A COPY OF THE TITLE INSURANCE POLICY, DEED TO THE PROPERTY OR SIGN AFFIDAVIT ATTESTING OWNERSHIP (see page 6)

D. WRITTEN SUBMITTAL REQUIREMENTS: (Section 3-913.A)

❏ Provide complete responses to the six (6) GENERAL APPLICABILITY CRITERIA – Explain **how** each criteria is achieved, in detail:

1. The proposed development of the land will be in harmony with the scale, bulk, coverage, density and character of adjacent properties in which it is located.

   _____
   _____
   _____

2. The proposed development will not hinder or discourage the appropriate development and use of adjacent land and buildings or significantly impair the value thereof.

   _____
   _____
   _____

3. The proposed development will not adversely affect the health or safety or persons residing or working in the neighborhood of the proposed use.

   _____
   _____
   _____

4. The proposed development is designed to minimize traffic congestion.

   _____
   _____
   _____

5. The proposed development is consistent with the community character of the immediate vicinity of the parcel proposed for development.

   _____
   _____
   _____

6. The design of the proposed development minimizes adverse effects, including visual, acoustic and olfactory and hours of operation impacts, on adjacent properties.

   _____
   _____
   _____
   _____

❏ Provide complete responses to the ten (10) COMPREHENSIVE INFILL REDEVELOPMENT PROJECT CRITERIA (as applicable) – Explain **how** each criteria is achieved in detail:

1. The development or redevelopment of the parcel proposed for development is otherwise impractical without deviations from the use, intensity and development standards.

   _____
   _____
   _____
   _____

FIGURE A2.18 Sample flexible development application. (continued)

2. The development of the parcel proposed for development as a comprehensive infill redevelopment project or residential infill project will not reduce the fair market value of abutting properties. (Include the existing value of the site and the proposed value of the site with the improvements.)

_____

_____

3. The uses within the comprehensive infill redevelopment project are otherwise permitted in the City of Clearwater.

_____

_____

4. The uses or mix of use within the comprehensive infill redevelopment project are compatible with adjacent land uses.

_____

_____

5. Suitable sites for development or redevelopment of the uses or mix of uses within the comprehensive infill redevelopment project are not otherwise available in the City of Clearwater.

_____

_____

6. The development of the parcel proposed for development as a comprehensive infill redevelopment project will upgrade the immediate vicinity of the parcel proposed for development.

_____

_____

7. The design of the proposed comprehensive infill redevelopment project creates a form and function that enhances the community character of the immediate vicinity of the parcel proposed for development and the City of Clearwater as a whole.

_____

_____

8. Flexibility in regard to lot width, required setbacks, height and off-street parking are justified by the benefits to community character and the immediate vicinity of the parcel proposed for development and the City of Clearwater as a whole.

_____

_____

9. Adequate off-street parking in the immediate vicinity according to the shared parking formula in Division 14 of Article 3 will be available to avoid on-street parking in the immediate vicinity of parcel proposed for development.

_____

_____

10. The design of all buildings complies with the Tourist District or Downtown District design guidelines in Division 5 of Article 3 (as applicable). Use separate sheets as necessary.

_____

_____

FIGURE A2.18 Sample flexible development application. (continued)

E. SUPPLEMENTAL SUBMITTAL REQUIREMENTS: (Code Section 4-202.A)

- ☐ SIGNED AND SEALED SURVEY (including legal description of property) – **One original and 12 copies**;
- ☐ TREE SURVEY (including existing trees on site and within 25' of the adjacent site, by species, size (DBH 4" or greater), and location, including drip lines and indicating trees to be removed);
- ☐ LOCATION MAP OF THE PROPERTY;
- ☐ PARKING DEMAND STUDY in conjunction with a request to make deviations to the parking standards (ie. Reduce number of spaces). Prior to the submittal of this application, the methodology of such study shall be approved by the Community Development Coordinator and shall be in accordance with accepted traffic engineering principles. The findings of the study will be used in determining whether or not deviations to the parking standards are approved;
- ☐ GRADING PLAN, as applicable;
- ☐ PRELIMINARY PLAT, as required (Note: Building permits will not be issued until evidence of recording a final plat is provided);
- ☐ COPY OF RECORDED PLAT, as applicable;

F. SITE PLAN SUBMITTAL REQUIREMENTS: (Section 4-202.A)

- ☐ SITE PLAN with the following information (not to exceed 24" x 36"):
  - ___ All dimensions;
  - ___ North arrow;
  - ___ Engineering bar scale (minimum scale one inch equals 50 feet), and date prepared;
  - ___ Location map;
  - ___ Index sheet referencing individual sheets included in package;
  - ___ Footprint and size of all EXISTING buildings and structures;
  - ___ Footprint and size of all PROPOSED buildings and structures;
  - ___ All required setbacks;
  - ___ All existing and proposed points of access;
  - ___ All required sight triangles;
  - ___ Identification of environmentally unique areas, such as watercourses, wetlands, tree masses, and specimen trees, including description and location of understory, ground cover vegetation and wildlife habitats, etc;
  - ___ Location of all public and private easements;
  - ___ Location of all street rights-of-way within and adjacent to the site;
  - ___ Location of existing public and private utilities, including fire hydrants, storm and sanitary sewer lines, manholes and lift stations, gas and water lines;
  - ___ All parking spaces, driveways, loading areas and vehicular use areas, including handicapped spaces;
  - ___ Depiction by shading or crosshatching of all required parking lot interior landscaped areas;
  - ___ Location of all solid waste containers, recycling or trash handling areas and outside mechanical equipment and all required screening {per Section 3-201(D)(i) and Index #701};
  - ___ Location of all landscape material;
  - ___ Location of all jurisdictional lines adjacent to wetlands;
  - ___ Location of all onsite and offsite storm-water management facilities;
  - ___ Location of all outdoor lighting fixtures; and
  - ___ Location of all existing and proposed sidewalks

- ☐ SITE DATA TABLE for existing, required, and proposed development, in written/tabular form:
  - ___ Land area in square feet and acres;
  - ___ Number of EXISTING dwelling units;
  - ___ Number of PROPOSED dwelling units;
  - ___ Gross floor area devoted to each use;
  - ___ Parking spaces: total number, presented in tabular form with the number of required spaces;
  - ___ Total paved area, including all paved parking spaces and driveways, expressed in square feet and percentage of the paved vehicular area;
  - ___ Size and species of all landscape material;
  - ___ Official records book and page numbers of all existing utility easement;
  - ___ Building and structure heights;
  - ___ Impermeable surface ratio (I.S.R.); and
  - ___ Floor area ratio (F.A.R.) for all nonresidential uses

- ☐ REDUCED SITE PLAN to scale (8 ½ X 11) and color rendering if possible

- ☐ FOR DEVELOPMENTS OVER ONE ACRE, provide the following additional information on site plan:
  - ___ One-foot contours or spot elevations on site;
  - ___ Offsite elevations if required to evaluate the proposed stormwater management for the parcel;
  - ___ All open space areas;
  - ___ Location of all earth or water retaining walls and earth berms;
  - ___ Lot lines and building lines (dimensioned);
  - ___ Streets and drives (dimensioned);
  - ___ Building and structural setbacks (dimensioned);
  - ___ Structural overhangs;
  - ___ Tree Inventory; prepared by a "certified arborist", of all trees 8" DBH or greater, reflecting size, canopy (drip lines) and condition of such trees

FIGURE A2.18 Sample flexible development application. (continued)

G. LANDSCAPING PLAN SUBMITTAL REQUIREMENTS: (Section 4-1102.A)

- ❑ LANDSCAPE PLAN:
  - ___ All existing and proposed structures;
  - ___ Names of abutting streets;
  - ___ Drainage and retention areas including swales, side slopes and bottom elevations;
  - ___ Delineation and dimensions of all required perimeter landscape buffers;
  - ___ Sight visibility triangles;
  - ___ Delineation and dimensions of all parking areas including landscaping islands and curbing;
  - ___ Proposed and required parking spaces;
  - ___ Existing trees on-site and immediately adjacent to the site, by species, size and locations, including dripline (as indicated on required tree survey);
  - ___ Location, size, description, specifications and quantities of all existing and proposed landscape materials, including botanical and common names;
  - ___ Typical planting details for trees, palms, shrubs and ground cover plants including instructions, soil mixes, backfilling, mulching and protective measures;
  - ___ Interior landscaping areas hatched and/or shaded and labeled and interior landscape coverage, expressing in both square feet and percentage covered;
  - ___ Conditions of a previous development approval (e.g. conditions imposed by the Community Development Board);
  - ___ Irrigation notes
- ❑ REDUCED LANDSCAPE PLAN to scale (8 ½ X 11) (color rendering if possible)
- ❑ IRRIGATION PLAN (required for Level Two and Three applications)
- ❑ COMPREHENSIVE LANDSCAPE PROGRAM application, as applicable

H. STORMWATER PLAN SUBMITTAL REQUIREMENTS: (City of Clearwater Design Criteria Manual and 4-202.A.21)

- ❑ STORMWATER PLAN including the following requirements:
  - ___ Existing topography extending 50 feet beyond all property lines;
  - ___ Proposed grading including finished floor elevations of all structures;
  - ___ All adjacent streets and municipal storm systems;
  - ___ Proposed stormwater detention/retention area including top of bank, toe of slope and outlet control structure;
  - ___ Stormwater calculations for attenuation and water quality;
  - ___ Signature of Florida registered Professional Engineer on all plans and calculations
- ❑ COPY OF PERMIT INQUIRY LETTER OR SOUTHWEST FLORIDA WATER MANAGEMENT DISTRICT (SWFWMD) PERMIT SUBMITTAL (SWFWMD approval is required prior to issuance of City Building Permit), if applicable
- ❑ COPY OF STATE AND COUNTY STORMWATER SYSTEM TIE-IN PERMIT APPLICATIONS, if applicable

I. BUILDING ELEVATION PLAN SUBMITTAL REQUIREMENTS: (Section 4-202.A.23)

Required in the event the application includes a development where design standards are in issue (e.g. Tourist and Downtown Districts) or as part of a Comprehensive Infill Redevelopment Project or a Residential Infill Project.

- ❑ BUILDING ELEVATION DRAWINGS – all sides of all buildings including height dimensions, colors and materials
- ❑ REDUCED BUILDING ELEVATIONS – four sides of building with colors and materials to scale (8 ½ X 11) (black and white and color rendering, if possible) as required

J. SIGNAGE: (Division 19. SIGNS / Section 3-1806)

- ❑ All EXISTING freestanding and attached signs; Provide photographs and dimensions (area, height, etc.), indicate whether they will be removed or to remain.
- ❑ All PROPOSED freestanding and attached signs; Provide details including location, size, height, colors, materials and drawing
- ❑ Comprehensive Sign Program application, as applicable (separate application and fee required).
- ❑ Reduced signage proposal (8 ½ X 11) (color), if submitting Comprehensive Sign Program application.

FIGURE A2.18 Sample flexible development application. (continued)

K. TRAFFIC IMPACT STUDY: (Section 4-202.A.13 and 4-801.C)

☐ Include as required if proposed development will degrade the acceptable level of service for any roadway as adopted in the Comprehensive Plan. Trip generation shall be based on the most recent edition of the Institute of Transportation Engineer's Trip General Manual. Refer to Section 4-801 C of the Community Development Code for exceptions to this requirement.

L. SIGNATURE:

I, the undersigned, acknowledge that all representations made in this application are true and accurate to the best of my knowledge and authorize City representatives to visit and photograph the property described in this application.

STATE OF FLORIDA, COUNTY OF PINELLAS
Sworn to and subscribed before me this _____ day of _____, A.D. 20____ to me and/or by _____, who is personally known has produced _____ as identification.

_____
Signature of property owner or representative

_____
Notary public,
My commission expires:

FIGURE A2.18 Sample flexible development application. (continued)

M. AFFIDAVIT TO AUTHORIZE AGENT:

_____
(Names of all property owners)

1. That (I am/we are) the owner(s) and record title holder(s) of the following described property (address or general location):

_____
_____

2. That this property constitutes the property for which a request for a: (describe request)

_____

3. That the undersigned (has/have) appointed and (does/do) appoint: _____

   as (his/their) agent(s) to execute any petitions or other documents necessary to affect such petition;

3. That this affidavit has been executed to induce the City of Clearwater, Florida to consider and act on the above described property;

4. That the applicant acknowledges that all impact fees (parks and recreation, traffic, etc.) will be paid PRIOR to the issuance of a building permit, certificate of occupancy, or other mechanism, whichever occurs first;

5. That site visits to the property are necessary by City representatives in order to process this application and the owner authorizes City representatives to visit and photograph the property described in this application;

6. That (I/we), the undersigned authority, hereby certify that the foregoing is true and correct.

_____
Property Owner

_____
Property Owner

STATE OF FLORIDA,
COUNTY OF PINELLAS

Before me the undersigned, an officer duly commissioned by the laws of the State of Florida, on this _____ day of _____, personally appeared _____ who having been first duly sworn Deposes and says that he/she fully understands the contents of the affidavit that he/she signed.

My Commission Expires:

_____
Notary Public

FIGURE A2.18 Sample flexible development application.

Appendix 2

PLACER COUNTY PLANNING DEPARTMENT

AUBURN OFFICE
11414 B Avenue
Auburn, CA 95603
530-886-3000/FAX 530-886-3080

TAHOE OFFICE
565 W. Lake Blvd./P. O. Box 1909
Tahoe City CA 96145
530-581-6280/FAX 530-581-6282

Web page: www.placer.ca.gov/planning   E-Mail : planning@placer.ca.gov

Reserved for Date Stamp

MAJOR SUBDIVISION APPLICATION

PURSUANT TO THE POLICY OF THE BOARD OF SUPERVISORS, THE PLANNING DEPARTMENT CANNOT ACCEPT APPLICATIONS ON TAX DELINQUENT PROPERTY. APPLICATIONS AFFECTING PROPERTIES WITH ZONING VIOLATIONS, OR OTHER VIOLATIONS OF COUNTY CODE, MAY BE REJECTED.

----TO BE COMPLETED BY THE APPLICANT----

1. Project Name _____ APN _____
2. Developer _____
   Address _____ Telephone Number _____ Fax Number _____
   _____ City _____ State _____ Zip Code _____
3. Engineer _____
   Address _____ Telephone Number _____ Fax Number _____
   _____ City _____ State _____ Zip Code _____
4. Total acreage _____ Number of proposed lots/units _____
   Proposed lot sizes: Minimum _____ Maximum _____ Average _____

_____
Signature of Applicant

INDEMNIFICATION AGREEMENT: I, the Subdivider, will defend, indemnify and hold harmless the County from any defense costs, including attorneys' fees or other loss connected with any legal challenge brought as a result of an approval concerning this Subdivision. I also agree to execute a formal agreement to this effect on a form provided by the County and available for my inspection.

SIGNATURE OF SUBDIVIDER _____

----OFFICE USE ONLY----

Date Tentative Map approved: _____ Expiration date _____
Date first extension approved: _____ New expiration date _____
Date second extension approved: _____ New expiration date _____
Auto. ext. of time per Sec. _____, Subdiv. Map Act New expiration date _____
Date last extension approved: _____ Final expiration date _____

POSTING OF PROPERTY: *At the time of application, posters will be provided by the Planning Department. These posters, in addition to notifying adjacent land owners of pending subdivision near their property, are used by county staff members to confirm they are looking at the correct piece of property when doing a field review. Should the staff members not be able to locate the property involved, the proposed subdivision will be continued to an open date by the Planning Commission until the required field review can be completed.*

FIGURE A2.19   Sample major subdivision application. (continued)

MAJOR SUBDIVISION

1. Submit one Initial Project Application;

2. Submit one Major Subdivision Application (Note: Application must include Indemnification Agreement signature); and

3. Submit a total of 15 tentative maps folded to 8½"x11", 1 (one) to be reduced to 8½" x 11". Maps should include information per Section 19.125 of the Land Development Manual. In addition:

    a. Accurately plot, label, and show exact location of the base and driplines of all protected trees (native 6" dbh or multi-trunk trees 10" dbh or greater) within 50 feet of any development activity (i.e. proposed structures, driveways, roadways, cuts/fills, underground utilities, lakes, recreation facilities, etc.) pursuant to Placer County Code, Chapter 36 (Tree Ordinance); and

    b. Provide an aerial photo of the site (1" = 100' or same scale as the proposed tentative map).

APPEALS - An appeal must be filed within 10 calendar days of the decision that is the subject of the appeal. An appeal application shall be submitted, along with the current filing fee, to the Planning Department. The appeal shall include any explanatory materials the appellant may wish to furnish. The Board of Supervisors will be the hearing body that will consider the appeal.

PUBLIC NOTICING REQUIREMENTS - The Planning Department shall notify all owners of property lying within 300 or more feet of the property, which is the subject of this project. In addition, the applicant shall post the property with posters furnished by the Planning Department at least 10 days prior to the scheduled hearing date (date and time will be available from the Planning Department approximately 20 days prior to the scheduled hearing.) One copy of the poster, together with the Affidavit of Posting, must be filed with the Planning Department prior to the hearing date.

FIGURE A2.19 Sample major subdivision application.

278 Appendix 2

PLACER COUNTY PLANNING DEPARTMENT

AUBURN OFFICE
11414 B Avenue
Auburn, CA 95603
530-886-3000/FAX 530-886-3080
Web page: www.placer.ca.gov/planning

TAHOE OFFICE
565 W. Lake Blvd./P. O. Box 1909
Tahoe City CA 96145
530-581-6280/FAX 530-581-6282
E-Mail : planning@placer.ca.gov

Reserved for Date Stamp

MINOR LAND DIVISION

Filing fee: _____ Receipt #_____ Hearing Date: _____ File #: **MLD-**_____

PURSUANT TO THE POLICY OF THE BOARD OF SUPERVISORS, THE PLANNING DEPARTMENT CANNOT ACCEPT APPLICATIONS ON TAX DELINQUENT PROPERTY. APPLICATIONS AFFECTING PROPERTIES WITH ZONING VIOLATIONS, OR OTHER VIOLATIONS OF COUNTY CODE, MAY BE REJECTED.

-----TO BE COMPLETED BY THE APPLICANT-----

The names listed below must be as they appear on the title to the properties because this application will be used to prepare the County's resolution of approval. If errors result from incorrect or incomplete information, the applicant will bear the cost of recording correcting documents.

1. Project name _____ APN:_____
2. Engineer (if any)_____
                                                 Telephone Number   Fax Number
       Address_____
                                     City   State   Zip Code
3. Size of property (acreage or square feet)_____
4. Date property purchased by present owner_____ Document number of where deed is recorded: _____
5. Number of parcels property is to be divided into:_____
       Parcel 1: _____sq.ft. or _____acres
       Parcel 2: _____sq.ft. or _____acres
       Parcel 3: _____sq.ft. or _____acres
       Parcel 4: _____sq.ft. or _____acres
       Remainder:_____sq.ft. or _____acres
6. Describe existing and proposed uses of the property:_____
7. Is this property or any portion thereof covered by a Williamson Act contract for the reduction of property taxes? Yes_____ No _____. If yes, provide contract number _____
8. Are parcels to be created as valid building sites? Yes _____ No _____. If no, explain: _____
   _____
9. Do you (applicant and owner) currently own any property, which is adjacent to that shown on your tentative map? Yes _____ No _____. If yes, show on tentative map, label as such, and be prepared to discuss in detail at the Parcel Review Committee meeting.
10. Have you (applicant and owner) ever owned any property adjacent to that shown on your tentative map? Yes _____ No _____. If yes, show on tentative map, label as such, and be prepared to discuss in detail at the Parcel Review Committee meeting.
11. Have you (applicant and owner) ever caused adjacent property to be divided either for yourself or for others by acting as applicant on previous divisions, through contractual arrangements or other

FIGURE A2.20   Sample minor land division application. (continued)

methods? Yes _____ No _____. If yes, be prepared to discuss in detail at the Parcel Review Committee meeting when this tentative map application will be considered.

12. Have you (applicant and owner) filed a "Notice of Intention" questionnaire and application for "Public Report" with the State Department of Real Estate? Yes _____ No _____. If no, do you intend to do so? Yes _____ No _____.

SPECIAL NOTE: If the applicant/owner or their representative is unable to provide details regarding items 9 through 12, this matter may be continued until the information is made available. If off-site access is used, then a guarantee of the access from a title company will be required. Access will have to meet the current requirements of County Code I hereby acknowledge that I have read this application and state that the information given is correct. I agree to comply with all County ordinances and State laws regulating property division.

_____   _____
Signature   Date

INSTRUCTIONS FOR FILING MINOR LAND DIVISION APPLICATION

1. Complete the application form and file along with the current filing fee; along with one copy of the Initial Project Application and one copy of the Exemption Verification form and filing fee.
2. Submit the appropriate number of tentative maps (5) that include the following information (per Section 19.310 of the Placer County Land Development Manual). The map shall be to scale and drawn clearly and legibly on one sheet of paper at least 8-1/2" x 11" in size. (If maps to be submitted are larger, please fold all copies to file size -- 8-1/2"x11".)
   Please note: At least 1 copy of this map must be 8-1/2" x 11"
   a. Names of property owner(s) and/or applicant of property.
   b. Boundary lines and taped or known dimensions of each parcel being created, using dashed lines.
   c. Proposed division lines with dimensions of each parcel being created, using dashed lines.
   d. All existing structures together with their dimensions and approximate distance from boundary lines.
   e. The approximate area of the original parcel and the minimum area of each new proposed parcel.
   f. Names, locations and widths of all existing traveled ways, including driveways, streets, and rights-of-way known to the owner as to location on or near the property.
   g. Approximate locations and widths of all new streets and rights-of-way proposed by the owner.
   h. Approximate location and dimensions of all existing and proposed easements, wells, leach lines, seepage pits or other underground structures.
   i. Approximate location and dimensions of all proposed easements for utilities and drainage.
   j. Approximate location of all creeks and drainage channels and a general indication of slope of the land.
   k. North arrow and approximate scale of drawing.
   l. Vicinity map which shows the property in relation to existing County roads and adjacent properties sufficient to identify it for field review and must be shown to the nearest 1/10th of a mile from County cross road.
   m. Accurately plot, label, and show exact location of the base and driplines of all protected trees (native trees 6" dbh or greater, or multi-trunk trees 10" dbh or greater) within 50 feet of any development activity (i.e., proposed structures, driveways, cuts/fills, underground utilities, etc.) pursuant to Placer County Code, Chapter 36 (Tree Ordinance). NOTE: A tree survey prepared by an I.S.A. certified arborist may be required. Verify with the Planning Department prior to submittal of this application.
   n. Site plan shall show all existing and proposed grading.

POSTING OF PROPERTY: At the time of application, posters will be provided by the Planning Department. These posters, in addition to notifying adjacent landowners of pending land divisions near their property, are used by County staff members to confirm they are looking at the correct piece of property when field reviewing proposed splits. Should the staff members not be able to locate the property involved in the Minor Land Division, the Parcel Review Committee will continue the proposed split to an open date until the required field review can be completed.

FIGURE A2.20   Sample minor land division application.

280 Appendix 2

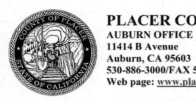

PLACER COUNTY PLANNING DEPARTMENT

Reserved for Date Stamp

AUBURN OFFICE	TAHOE OFFICE
11414 B Avenue	565 W. Lake Blvd./P. O. Box 1909
Auburn, CA 95603	Tahoe City CA 96145
530-886-3000/FAX 530-886-3080	530-581-6280/FAX 530-581-6282
Web page: www.placer.ca.gov/planning	E-Mail : planning@placer.ca.gov

DESIGN/SITE REVIEW APPLICATION

Filing fee: _____ Receipt # _____ File No.: DSA- _____

THE DESIGN REVIEW IS VALID FOR 24 MONTHS (UNLESS IT IS STATED OTHERWISE) OR IT IS EXERCISED BY ACTUAL CONSTRUCTION ONSITE. EXTENSIONS OF TIME MAY BE GRANTED FOR NO MORE THAN A TOTAL OF THREE YEARS AS PROVIDED BY SECTION 20.160(C) OF THE ZONING ORDINANCE. APPLICANT MUST APPLY PRIOR TO THE EXPIRATION DATE.

-----TO BE COMPLETED BY THE APPLICANT-----

1. Name of project _____ Assessor's Parcel Number _____
2. Applicant's name and address _____

   Telephone: _____ Fax: _____ E-Mail: _____
3. Contractor's name and address _____

   Telephone: _____ Fax: _____ E-Mail: _____
4. Landscape designer's name and address _____

   Telephone: _____ Fax: _____ E-Mail: _____
5. Architect's name and address _____

   Telephone: _____ Fax: _____ E-Mail: _____
6. Development proposed (include uses, building size, improvements, remodel, etc) _____

7. Number of required parking spaces _____ Number of proposed parking spaces _____

_____
Signature of Applicant

FIGURE A2.21   Sample design/site review application. (continued)

To expedite issuance of a Design Review approval, the applicant is required to submit the following along with one copy of the Initial Project Application and Exemption Verification form: (All maps shall be to scale & folded to no larger than 8-1/2" x 11")

Ten (10) copies of the site plan which shows the following: (* Tahoe applications - submit 12 copies of map submittals #1-4)

- Lot location (name of roads, distance from and name of nearest intersecting road, landmarks, etc.);
- Lot dimensions, driveway width, parking space size and building size;
- Location of and use of existing and proposed structures on the property;
- Setbacks from property lines of all buildings, signs, fences, etc;
- North arrow and scale;
- Existing and proposed contours;
- Existing and proposed grading;
- Methods of access to nearest road;
- Pavement widths;
- Off-street parking design;
- Transformer locations;
- Easements;
- Accurately plot, label and show exact location of the base and driplines of all protected trees (native trees 6" dbh or greater, or multi-trunk trees 10" dbh or greater) within 50' of any development activity (e.g.: proposed structures, driveways, cuts/fills, underground utilities, etc.) pursuant to Placer County Code, Chapter 36 (Tree Ordinance).

NOTE: A tree survey prepared by an I.S.A. certified arborist may be required. Verify with the Planning Department prior to submittal of this application.

- Existing rock outcrops, riparian areas or other natural features;
- Location of freestanding lights; and
- Building envelopes.
- Ten (10) copies of the exterior building elevations of all sides of proposed building. Show proposed and/or existing exterior finish of all structures, including all colors proposed for trim, walls, roofing, windows, doors, mechanical equipment, lights, etc.
- Ten (10) copies of proposed fences and walls (show height and material(s). Include trash enclosure location(s), materials, and height (may be included in elevation drawings).
- Ten (10) copies of the landscape plan (see Landscape Design Guidelines)—include size, planting notes, botanical and common names, spacing and number of all plant materials. A registered landscape architect or an approved landscape designer should prepare all landscape plans.
- Vicinity map which shows the location of the subject property in relation to existing County roads and adjacent properties sufficient to identify the property in the field for someone unfamiliar with the area. The distance to the closest intersection of County roads should be shown to the nearest 1/10th of a mile.
- In addition to the above information, the following is typically required prior to Design Review approval:
- Complete irrigation plans.
- Exterior lighting – for freestanding lights, indicate location, height, wattage, type of fixture and materials; for building lights indicate the location and type of fixture.
- Size, location, style, colors, materials and type of illumination of all signs existing and proposed on the property.
- Approval by P.G. & E. of Site Plan showing transformer location (Item # 1 [k] above).
- The relationship of proposed building(s) to all other structures within 100 feet and their height.
- Roof Plan showing roof slope & materials, general size and location of all mechanical equipment and vents, ducts, and other roof mounted items.
- Other pertinent information as required by the Design/Site Review Committee.

If any of this information is not available or will be developed at a later date, please inquire at the Planning Department about a later submittal of such information.

FIGURE A2.21 Sample design/site review application.

PLACER COUNTY PLANNING DEPARTMENT

AUBURN OFFICE
11414 B Avenue
Auburn, CA 95603
530-886-3000/FAX 530-886-3080
Website: www.placer.ca.gov/planning

TAHOE OFFICE
565 W. Lake Blvd./P. O. Box 1909
Tahoe City CA 96145
530-581-6280/FAX 530-581-6282
E-Mail: planning@placer.ca.gov

Reserved for Date Stamp

CONDITIONAL USE PERMIT/MINOR USE PERMIT

Filing fee: $_____ Type:_____ Receipt #_____ File # CUP-_____ MUP-_____

Hearing Date _____

--TO BE COMPLETED BY THE APPLICANT--

1. Project Name _____
2. Applicant _____
3. Project Description _____

PLEASE SUBMIT A WELL-DETAILED SITE PLAN (see instructions for requirements)

4. Assessor's Parcel Number(s) _____

SIGNATURE OF APPLICANT:_____

INDEMNIFICATION AGREEMENT: *I, the Applicant, will defend, indemnify, and hold harmless the County from any defense costs, including attorneys' fees or other loss connected with any legal challenge brought as a result of an approval concerning this Entitlement. I also agree to execute a formal agreement to this effect on a form provided by the County and available for my inspection.*

SIGNATURE OF APPLICANT:_____

PERMITS GRANTED FOR AN INDEFINITE PERIOD AUTOMATICALLY *EXPIRE 24 MONTHS AFTER DATE OF ISSUANCE* IF NOT EXERCISED BY THAT TIME, AS PROVIDED BY SECTION 17.58.160(B)(1) OF THE PLACER COUNTY ZONING ORDINANCE.

--OFFICE USE ONLY--

DECISION OF HEARING BODY: On _____, the Planning Commission/Zoning Administrator approved/denied this application subject to the attached list of _____ findings/conditions.

_____

--FOR USE AFTER PUBLIC HEARING--

I have read the above/attached conditions and will comply:

SIGNATURE OF APPLICANT:_____

PLEASE RETURN ONE SIGNED COPY

FIGURE A2.22   Sample conditional use permit/minor use permit application. (continued)

Use Permits (Minor or Conditional) shall only be approved subject to the findings as noted in Section 17.58.140(A) of the Zoning Ordinance. In conditionally approving a Minor or Conditional Use Permit, the granting authority shall adopt conditions of approval as necessary to accomplish the objectives as set forth in Section 17.58.140(B) of the Zoning Ordinance.

Complete an Initial Project Application, an Exemption Verification form and the Use Permit application (including Indemnification Agreement signature). Submit the applications and filing fee, along with **15 site plans** drawn to an acceptable scale, no larger than 8½" x 11" (or **folded** to that size). **Site plans** shall contain the following information:

1. Boundary lines and dimensions of parcel(s);
2. Existing and proposed structures and their gross floor area in square feet, parking areas with spaces delineated, distance between structures and the distance from property lines;
3. The approximate area of the parcel, in square feet or acres;
4. Names, locations and widths of all existing traveled ways, including driveways, streets and right-of-ways on or adjacent to the property;
5. Approximate locations and widths of all proposed streets, right-of-ways, driveways and/or parking areas;
6. Approximate location and dimensions of all existing easements, well, leach lines, seepage pits or other underground structures;
7. Approximate location and dimensions of all proposed easements for utilities and drainage.
8. Approximate location of all creeks and drainage channels and a general indication of the slope of the land and all trees of significant size;
9. Accurately plot, label and show existing locations of the base and driplines of all protected trees (Native trees 6" dbh or greater or multi-trunk trees 10" dbh or greater) within 50 feet of any development activity (i.e.: proposed structures, driveways, cuts/fills, underground utilities, etc.) pursuant to Placer County Code, Chapter 12 (Tree Ordinance). NOTE: A tree survey prepared by and I.S.A. certified arborist may be required. Verify with the Planning Department prior to submittal of this application;
10. Show all existing and proposed grading;
11. North arrow and scale of drawing;
12. **Vicinity map**, which shows the location of the subject property in relation to existing County roads and adjacent properties, sufficient to identify the property in the field to someone unfamiliar with the area. The distance to the closest intersection of County roads should be shown to the nearest $1/10^{th}$ of a mile;
13. Assessor's parcel number;
14. Name(s) of the project property owner(s) and applicant.

Where the proposed project includes the construction of a building(s), preliminary elevations should be provided in order to assist the staff and hearing body in reviewing the proposed project. (5 copies for Zoning Administrator items, and 15 for Planning Commission items.)

NOTE: 15 copies of the site plan are required - no more than 8½"x11" or **folded** to that size unless the project planner or hearing clerk determines additional copies are required.

APPEALS - An appeal must be filed within 10 calendar days of the decision that is the subject of the appeal. An appeal application shall be submitted, along with the current filing fee, to the Planning Department. The appeal shall include any explanatory materials the appellant may wish to furnish. The Planning Commission or Board of Supervisors will be the hearing body that will consider the appeal (based on the type of Use Permit involved).

Prior to the commencement of any grading and/or construction activities on the property in question, that are based upon the entitlements conferred by Placer County permit approval(s), the applicant should consult with the California Department of Fish & Game (DFG) to determine whether or not a Streambed Alteration Agreement [§1603, CA Fish & Game Code] is required. The applicant should also consult with the U.S. Army Corps of Engineers to determine whether or not a permit is required for these activities pursuant to Section 404 of the Clean Water Act. **The applicant's signature on this application form signifies an acknowledgement that this statement has been read and understood.**

FIGURE A2.22 Sample conditional use permit/minor use permit application.

Appendix 2

PERMIT NO: _____ **RECEIPT NO:** _____ **CHECK NO:** _____ **PERMIT AMT:** _____

OFFICE OF PLANNING
PRINCE WILLIAM COUNTY, VIRGINIA

UNDERGROUND UTILITY LINE PERMIT
(Telephone, Electric, Gas, & Cable)

PLEASE DO NOT WRITE ABOVE THIS LINE

Plan Name: _____ Plan No.: _____

Owner/Developer's Name: _____

Utility Company's Name: _____

Utility Company's Address: _____

Telephone No: _____ Address of Site Location: _____

Plan Approval Date: _____ Plan Expiration Date: _____

GPIN: _____ Magisterial District: _____

Date Grading/Site Development/Site Preparation Permit Issued: _____

Permission is hereby given to the above utility company insofar as Prince William County has the right and power to grant the same to perform the construction as shown on the approved utility installation plan(s). Said work to be completed in a manner satisfactory to Prince William County by _____.
(Permit Expiration Date - 12 Months From Date of Issuance)

Prince William County reserves full governmental control over the subject matter of this utility permit. Utilities shall be installed underground in accordance with standards of utility practice for underground construction, and in accordance with standards furnished to and regulations issued by any applicable regulatory authority. Such standards, and any amendments thereto, shall be furnished to the county by the utility company, and shall comport with the guidelines specified in the land use permit manual of the Virginia Department of Transportation, Section 3.200 titles "Guidelines for the Accommodation of the Utility Facilities within the Right-of-Way of Highways". Utility line installation is expected to conform to all the regulations of the Federal Occupational Safety and Health Administration and the Virginia Safety and Health Codes Commission. Should the utility company deviate from the approved plans, without prior approval from Prince William County, this permit shall be considered void and of no effect.

The signatures at the bottom of this permit, indicating approval of the installed erosion control devices by the Watershed Management Division Site Inspector is required before proceeding with land disturbing(construction) activities. This permit shall be considered null and void if this requirement is not met. All erosion control devices, inclusive of dust control, must be satisfactorily maintained until the final release of siltation and erosion control escrow. Mud tracking out of the project site is prohibited.

Application is hereby made for a permit as indicated above and shown on the accompanying plan or sketch. Said work will be done in compliance with the provisions, rules and regulations of the Virginia Department of Transportation, Virginia Erosion Control Law, the Code of Prince William County, the State Corporation Commission of Virginia, the Office of Pipeline Safety of the U.S. Department of Transportation, or the U.S. Department of Labor, so far as said rules are applicable thereto. Upon completion of construction, applicant agrees to maintain all improvements as imposed by the above described entities.

For utility installation in subdivisions where the streets are not yet in the State System, the utility company shall secure approval for access and construction from the developer (Prince William County Design and Construction Standards Manual).

If work is performed in an existing State right-of-way, (road that has been assigned a route number and is maintained by the State), a separate permit is required from the Virginia Department of Transportation. This permit does not authorize construction within privately owned rights-of-way or property.

The owner must notify the Department of Public Works, Watershed Management Division at 703-792-7070 at least 24 hours prior to the start of construction with applicable County ordinances and policies. THIS PERMIT MUST BE KEPT AT THE PROJECT SITE AND SHOWN WHEN REQUESTED.

· I have read all statements on this permit, understand the meaning, and hereby agree to abide by the provisions of this permit.

Signature: _____ Date: _____
Utility Company's/Developer's Signature

_____
Print Name and Title

Permit Issued By: _____ Date: _____
Signature/Agent for Prince William County

Erosion Control Devices Approved By: _____ Date: _____
Watershed Management Division Site Inspector

DC: Department of Public Works/Plan File/Permit File

FIGURE A2.23 Sample underground utility line permit.

CITY OF CLEARWATER
APPLICATION FOR A CLEARING AND/OR GRUBBING PERMIT
PLANNING & DEVELOPMENT SERVICES
MUNICIPAL SERVICES BUILDING, 100 SOUTH MYRTLE AVENUE, 2nd FLOOR
PHONE (727) 562-4567   FAX (727) 562-4576

APPLICANT NAME _____

MAILING ADDRESS: _____

PHONE NUMBER: _____ FAX NUMBER: _____

PROPERTY OWNERS: _____
(List all owners)

STREET ADDRESS OF
CLEARING/GRUBBING: _____
(IF DIFFERENT FROM APPLICANT'S ADDRESS)

PARCEL NUMBER: _____

PURPOSE OF CLEARING
AND/OR GRUBBING: _____

TYPES OF EQUIPMENT
TO BE USED: _____

DATE OF COMMENCEMENT: _____

DATE OF COMPLETION: _____

METHOD OF DEBRIS
DISPOSAL: _____

ALL APPLICATIONS FOR CLEARING AND/OR GRUBBING MUST BE ACCOMPANIED BY THE FOLLOWING INFORMATION:
___ SCALED SITE PLAN OR SCALED AERIAL PHOTO SHOWING PROPERTY BOUNDARIES, PHYSICAL AND NATURAL FEATURES AND LIMITS OF THE PROPOSED WORK.
___ COPY OF LEVEL 1 OR LEVEL 2 APPROVAL.
___ METHODS OF SOIL EROSION AND SEDIMENTATION CONTROL TO BE UNDERTAKEN DURING EARTHWORK ACTIVITIES AND THE MEANS AND TIMING OF SOIL STABILIZATION SUBSEQUENT TO THE COMPLETION OF THE CLEARING AND GRUBBING ACTIVITIES.

FIGURE A2.24 Sample application for a clearing and/or grubbing permit. (continued)

Appendix 2

I, the undersigned, acknowledge that all representations made in this application are true and accurate to the best of my knowledge.

STATE OF FLORIDA, COUNTY OF PINELLAS
Sworn to and subscribed before me this _____ day of _____, A.D., 19_____ to me and/or by _____, who is personally known has produced _____ as identification.

_____
Signature of property owner or representative

_____
Notary public,
my commission expires:

S: application forms/development review/land clearing and grubbing permit.doc

FIGURE A2.24 Sample application for a clearing and/or grubbing permit. (continued)

✓ "What I Need To Do" Checklist
This document is not a Permit   My Permit Application# :_____

General – *Please read the following completely before contacting review agents*:
- It is the Applicant's responsibility to actively pursue all approvals for a Land Use Permits(LUP).
- All releases from required approval agencies must be received at Division of Permits & Compliance(P&C) prior to permit issuance.
- The issued Land Use Permit will be mailed to the Applicant as it appears on Applicant's Address.
- Application(s) will not be forwarded to review agents until the application fee is paid in full.
- **No construction** may begin until the Applicant receives an issued Land Use Permit.
- Permit and plan status inquiries are encouraged on a weekly basis.
- The *Processing Fee* of $_____ is **non refundable**, even if the permit is canceled or denied.
- If you have questions or need help, please call the Permits office. Additional conferences will be arranged at your convenience.

FIRST: After application, **I will** :   [x] post the yellow location marker   [x] mark property lines
                                              [x] flag corners of proposed structure   [ ] mark proposed driveway entrance

I will also need to provide the following, these documents or actions are **required** to process my application.
☐ Variance &/or Special Exception from the Board of Zoning Appeals. The deadline for the _____ hearing is _____.
☐ Floodplain Management - Please submit _____ sets of complete plans to the Permits office for distribution and review.
☐ Site Development Plans - Please submit _____ sets of complete plans to the Permits office for distribution and review.
☐ Building/Construction Plans - Please submit _____ sets of complete plans to the Permits office for distribution and review
☐ Notify the Permits Office of the qualified ☐electrical ☐plumbing inspection agency **chosen by me** or my agent to perform inspections on the project. Upon notification, the Permits office will forward a copy of my application to this agency. My agency **may** contact Permits for me.
☐ Notify the Permits Office of the qualified master ☐electrician ☐plumber, including MD licenses numbers, I have contracted.

NEXT: In two(2) working days, i.e._____, I will contact by telephone the following agencies marked with a '✓' or 'X' to schedule inspections, field meetings or discuss these agencies requirements(i.e. plan requirements, bonds, etc.). By the aforementioned date, the following agency(ies) are scheduled to have a processed copy of my application making them aware of what I plan to do.
☐ Allegany County Planning Division for street name/structure address at **301-777-3093** (Matt Diaz - Ext 290).
☐ Soil Conservation District at **301-777-1747** for sediment and stormwater management plans (Bernie Connor Ext 109).
☐ ACDPW County Roads Division at **301-777-5955** for driveway inspection/bond submittal.
☐ State Highway Administration for inspections/bond submittal for: ☐ residential driveways at **301-729-8433**.
                                                                    ☐ commercial entrances at **301-729-8465** (J. Wolford).
                                                                    ☐ signs and billboards at **301-729-8451** (M.Murphy).
☐ Maryland Department of the Environment at **301-689-8598** for Waterway Construction Permit (Sean McKewen)
☐ LaVale Zoning Board at **301-724-2285** for zoning questions(*answering service*). I will need to attend a LaVale Zoning Board hearing. These hearings are generally conducted on the 2nd & 4th Mondays of the month at 7:00 p.m. within the lower LaVale Fire Hall. Sketch plan required
☐ LaVale Sanitary Commission at **301-729-1638** for ☐sewer and/or ☐water tap
☐ Other_____@_____

NOTE: *The following agencies marked with a '✓', '●' or 'X', will notify me of actions necessary for permit issuance.*
Division of Permits and Compliance for Major/Minor Subdivision requirements, review and approval-if necessary.(Dave Dorsey).
Division of Permits and Compliance for zoning certifications.
Allegany County Public Utilities Division for ☐sewer and/or ☐water tap (bill will be mailed to Applicant by Utilities Division).
Al Co Health Dept for (1)well/septic/tap permits,(2)buffers,(3)food service permits, (4)subdivision plats,(5)other requirements.
Maryland State Fire Marshal Office for review and approval (Hagerstown, MD Office).
ACDPW Engineering Division for ☐subdivision plats / ☐stormwater management plan review and approval.

MEANWHILE: *If I have questions, I can contact the Permits office for:*
Permit status inquiries or assistance	(301)777-5951 ext. #295
Zoning Certifications	(301)777-5951 ext. #293 or 292
Subdivision Plats	(301)777-5951 ext. #292
Floodplain Management	(301)777-5951 ext. #293
Stormwater and Erosion Control Plans	(301)777-5951 ext. #293
Building Code questions or appeals	(301)777-5951 ext. #352

_____ Make check payable to: **Allegany County Tax & Utility Office**
                                County Office Complex
                                701 Kelly Road
                                Cumberland, MD 21502-3401

_____ Mail permit materials, check & invoice to: **Division of Permits & Compliance**
                                                    County Office Complex Suite 112
                                                    701 Kelly Road
                                                    Cumberland, MD 21502-3401

_____ P&C's fax# is: 301-777-5950   24hrs/7days week

\\archive documents\forms\permit_ckecklist._2.doc
rev 8/00   [E3110!] /js

Applicant's Acknowledgment
The following was received and explained:
☐ "What I Need To Do" Checklist
☐ yellow location marker(triangular, paper flag)
☐ stakes
☐ Invoice #
☐ Major Site Plan Development Standards
☐ Inspection Agencies List
☐ Building Code Requirements
☐

Applicant _____ Date _____

FIGURE A2.25   Sample "What I Need To Do" checklist.

Your Company Name
Your Company Address
Your Company Phone and Fax Numbers

QUOTE

This agreement, made this _____ day of _____, 19__, shall set forth the whole agreement, in its entirety, by and between Your Company Name, herein called Contractor, and _____, herein called Owners.

Job name: _____

Job location: _____

The Contractor and Owners agree to the following: Contractor shall perform all work as described below and provide all material to complete the work described below. Contractor shall supply all labor and material to complete the work according to the attached plans and specifications. The work shall include the following: _____
_____
_____
_____

SCHEDULE

The work described above shall begin within _____ days of notice from Owners, with an estimated start date of _____. The Contractor shall complete the above work in a professional and expedient manner within _____ days from the start date.

PAYMENT SCHEDULE

Payment shall be made as follows: _____
_____
_____

This agreement, entered into on _____, shall constitute the whole agreement between Contractor and Owners.

_____   _____
Contractor   Date   Owner
Date

                                      _____
                                      Owner   Date

FIGURE A2.26 Sample quote form.

Your Company Name
Your Company Address
Your Company Phone and Fax Numbers

PROPOSAL

Date: _____

Customer name: _____
Address: _____
Phone number: _____
Job location: _____

DESCRIPTION OF WORK

Your Company Name will supply, and or coordinate, all labor and material for the above referenced job as follows: _____
_____
_____
_____
_____
_____
_____

PAYMENT SCHEDULE

Price: _____ dollars ($_____)

Payments to be made as follows: _____
_____
_____

All payments shall be made in full, upon presentation of each completed invoice. If payment is not made according to the terms above, Your Company Name will have the following rights and remedies. Your Company Name may charge a monthly service charge of _____ (_____%) percent, _____ (_____%) percent per year, from the first day default is made. Your Company Name may lien the property where the work has been done. Your Company Name may use all legal methods in the collection of monies owed to it. Your Company Name may seek compensation, at the rate of $_____ per hour, for attempts made to collect unpaid monies.

(Page 1 of 2. Please initial _____.)

FIGURE A2.27 Sample proposal form. (continued)

Your Company Name may seek payment for legal fees and other costs of collection, to the full extent the law allows.

If the job is not ready for the service or materials requested, as scheduled, and the delay is not due to Your Company Name's actions, Your Company Name may charge the customer for lost time. This charge will be at a rate of $_____ per hour, per man, including travel time.

If you have any questions or don't understand this proposal, seek professional advice. Upon acceptance, this proposal becomes a binding contract between both parties.

Respectfully submitted,

Your Name
Title

ACCEPTANCE

We the undersigned do hereby agree to, and accept, all the terms and conditions of this proposal. We fully understand the terms and conditions, and hereby consent to enter into this contract.

Your Company Name   Customer

By: _____   _____

Title: _____   Date: _____

Date: _____

Proposal expires in 30 days, if not accepted by all parties.

FIGURE A2.27 Sample proposal form.

Form RD 442-3
(Rev. 3-97)

Position 3

Name

Address

FORM APPROVED
OMB No. 0575-0015

BALANCE SHEET

	Month Day Year Current Year	Month Day Year Prior Year

ASSETS
CURRENT ASSETS
1. Cash on hand in Banks ....
2. Time deposits and short-term investments ....
3. Accounts receiveable ....
4.   Less: Allowance for doubtful accounts .... ( ) ( )
5. Inventories ....
6. Prepayments ....
7. _____ ....
8. _____ ....
9.   Total Current Assets *(Add 1 through 8)* ....

FIXED ASSETS
10. Land ....
11. Buildings ....
12. Furniture and equipment ....
13. _____ ....
14.   Less: Accumulated depreciation .... ( ) ( )
15. Net Total Fixed Assets *(Add 10 through 14)* ....

OTHER ASSETS
16. _____ ....
17. _____ ....
18. Total Assets *(Add 9, 15, 16 and 17)* ....

LIABILITIES AND EQUITIES
CURRENT LIABILITIES
19. Accounts payable ....
20. Notes payable ....
21. Current portion of USDA note ....
22. Customer deposits ....
23. Taxes payable ....
24. Interest payable ....
25. _____ ....
26. _____ ....
27.   Total Current Liabilities *(Add 19 through 26)* ...

LONG-TERM LIABILITIES
28. Notes payable USDA ....
29. _____ ....
30. _____ ....
31. Total Long-Term Liabilities *(Add 28 through 30)* ....
32. Total Liabilities *(Add 27 and 31)* ....

EQUITY
33. Retained earnings ....
34. Memberships ....
35. Total Equity *(Add lines 33 and 34)* ....
36. Total Liabilities and Equity *(Add lines 32 and 35)* ..

CERTIFIED CORRECT   Date   Appropriate Official *(Signature)*

According to the Paperwork Reduction Act of 1995, no persons are required to respond to a collection of information unless it displays a valid OMB control number. The valid OMB control number for this information collection is 0570-0015. The time required to complete this information is estimated to average 1 hour per response, including the time for reviewing instructions, searching existing data sources, gathering and maintaining the data needed, and completing and reviewing the collection of information.

RD 442-3 (Rev. 3-97)

FIGURE A2.28   Sample balance sheet. (continued)

INSTRUCTIONS

Present Borrowers

This form may be used as a year end Balance Sheet by Rural Development Community Program and Farm Service Agency Group Farm Loan Program borrowers who do not have an independent audit. Submit two copies within 60 days following year's end to the Agency Official. An independently audited balance sheet will substitute for this form.

Applicants

In preparing this form when the application for financing is for a facility which is a unit of your overall operation, two balance sheets are to be submitted: one for the facility being financed and one for the entire operation. Examples: (a) application to finance a sewage system which is a part of a water-sewage system or municipality, (b) application to finance a nursing home which is part of a larger health care facility.

Preparation of this Form

1. Enter data where appropriate for the current and prior year.

2. Line 35, Total Equity, of this form will be the same as line 26, on Form RD 442-2, "Statement of Budget, Income and Equity", when using the form.

3. The term Equity is used interchangeably with Net Worth, Fund Balance, etc.

BALANCE SHEET ITEMS

Current Assets

1. Cash on hand and in Banks
   Includes undeposited cash and demand deposits.

2. Time Deposits and Short Term Investments
   Funds in savings accounts and certificates of deposit maturing within one year.

3. Accounts Receivable
   Amounts billed but not paid by customers, users, etc. This is the gross amount before any allowances in item 4.

4. Allowance for Doubtful Accounts
   Amounts included in item 3 which are estimated to be uncollectible.

5. Inventories
   The total of all materials, supplies and finished goods on hand.

6. Prepayments
   Payments made in advance of receipt of goods or utilization of services. Examples: rent, insurance.

7 - 8. List other current assets not included above.

Fixed Assets

10 - 12. List land, buildings, furniture and equipment separately by gross value.

13. List other fixed assets.

14. Accumulated Depreciation
Indicate total accumulated depreciation for items 10-13.

Other Assets

16 - 17. List other assets not previously accounted for.

Current Liabilities

19. Accounts Payable
    Amounts due to creditors for goods delivered or services completed.

20. Notes Payable
    Amounts due to banks and other creditors for which a promissory note has been signed.

21. Current Portion USDA Note
    Amount due USDA for principal payment during the next 12 months. Includes any payments which are in arrears.

22. Customer Deposits
    Funds of various kinds held for others.

23. Taxes Payable

24. Interest Payable USDA
    Interest applicable to principal amount in line 21.

25 - 26. List other payables and accruals not shown above.

Long Term Liabilities

28. Notes Payable USDA
    List total principal payments to USDA which mature after one year and are not included in line 21.

29 - 30. List all other long term liabilities such as bonds, bank loans, etc. which are due after one year.

Equity

33. Retained Earnings
    Net income which has been accumulated from the beginning of the operation and not distributed to members, users, etc.

34. Memberships
    The total of funds collected from persons of membership type facilities, i.e., water and sewer systems.

FIGURE A2.28 Sample balance sheet. (continued)

FORMS MANUAL INSERT

FORM RD 442-3

Used by
Rural Development
Community Program
and Farm Service
Agency Group Farm
Loan Program
applicants and
borrowers.

(see reverse)

PROCEDURE FOR PREPARATION : RD Instructions 1942-A, 1951-E and 1955-A.

PREPARED BY : Applicant/Borrower.

NUMBER OF COPIES : Applicant - Original and one copy.
Borrower - Original and three copies.

SIGNATURES REQUIRED : Appropriate Applicant/Borrower Official.

DISTRIBUTION OF COPIES : Applicant - Original to County case docket; copy retained by Applicant. Borrower - Original and two copies to County; copy retained by Borrower; original to case docket, two copies to State Office (for Community Program delinquent Borrowers, State Office will sent copy to National Office).

FIGURE A2.28 Sample balance sheet.

Appendix 2

USDA-RD
Form RD 442-7
(Rev. 3-02)

Position 3
OPERATING BUDGET

Form Approved
OMB No. 0575-0015

Schedule 1

Name		Address		
Applicant Fiscal Year		County		State *(Including ZIP Code)*
From	To			

	20___ (1)	20___ (2)	20___ (3)	20___ (4)	First Full Year (5)
OPERATING INCOME					
1. _____					
2. _____					
3. _____					
4. _____					
5. Miscellaneous					
6. Less: Allowances and Deductions	( )	( )	( )	( )	( )
7. Total Operating Income *(Add Lines 1 through 6)*					
OPERATING EXPENSES					
8. _____					
9. _____					
10. _____					
11. _____					
12. _____					
13. _____					
14. _____					
15. Interest *(RD)*					
16. Depreciation					
17. Total Operating Expense *(Add Lines 8 through 16)*					
18. NET OPERATING INCOME *(LOSS) (Line 7 less 17)*					
NONOPERATING INCOME					
19. _____					
20. _____					
21. Total Nonoperating Income *(Add Lines 19 and 20)*					
22. NET INCOME *(LOSS)* *(Add Lines 18 and 21)* *(Transfer to Line A Schedule 2)*					

Budget and Projected Cash Flow Approved by Governing Body

Attest: _____
Secretary *Date*

_____
Appropriate Official *Date*

According to the Paperwork Reduction Act of 1995, an agency may not conduct or sponsor, and a person is not required to respond to a collection of information unless it displays a valid OMB control number. The valid OMB control number for this information collection is 0575-0015. The time required to complete this information collection is estimated to average 5 hours per response, including the time for reviewing instructions, searching existing data sources, gathering and maintaining the data needed, and completing and reviewing the collection of information.

FIGURE A2.29 Sample operating budget. (continued)

PROJECTED CASH FLOW

Schedule 2

	20___	20___	20___	20___	First Full Year
A. Line 22 from Schedule 1 Income *(Loss)*					
Add					
B. Items in Operations not Requiring Cash:					
1. Depreciation *(Line 16, Schedule 1)*					
2. Others: _____					
C. Cash Provided from:					
1. Proceeds from RD loan/grant					
2. Proceeds from others					
3. Increase *(Decrease)* in Accounts Payable, Accurals and other Current Liabilities					
4. Decrease *(Increase)* in Accounts Receivable, Inventories and Other Current Assets *(Exclude Cash)*					
5. Other: _____					
6. _____					
D. Total all A, B, and C Items					
E. *Less:* Cash Expended for:					
1. All Construction, Equipment and New Capital Items *(Loan and grant funds)*					
2. Replacement and Additions to Existing Property, Plant and Equipment					
3. Principal Payment RD Loan					
4. Principal Payment Other Loans					
5. Other: _____					
6. Total E 1 through 5					
Add					
F. Beginning Cash Balances					
G. Ending Cash Balances *(Total of D minus E 6 plus F)*					
Item G Cash Balance Composed of:					
Construction Account	$	$	$	$	$
Revenue Account					
Debt Payment Account	$	$	$	$	$
O&M Account					
Reserve Account					
Funded Depreciation Account					
Others: _____					
Total - Agrees with Item G	$	$	$	$	$

FIGURE A2.29 Sample operating budget. (continued)

Instructions - Operating Budget Schedule 1

This form is to be prepared by the Applicant and is to include data for each year, from loan closing through the first full year of operation. Example: If only two columns are required, use columns four(4) and five(5).

Income and Expense Items:

All data entered should be on the same basis as the Applicant's Accounting records, i.e., cash basis, accrual basis, etc.

Operating Income:

lines 1-5	List types of income as appropriate
line 6 —	Allowances and Deductions
	(Pertains generally to Health Care Institutions, and represents the difference between Gross Income and Amounts Received or to be Received from patients and third party payors)

Operating Expenses:

lines 8-14	List types of expenses as appropriate
line 15 —	Interest RD
	(Interest expense incurred on RD note(s))
line 16 —	Depreciation
	(Total depreciation expense for the year)
line 18 —	Net Operating (Loss)
	(This amount represents the net operating income or loss before adding income not related to operations below)

Non Operating Income:

lines 19-20	Indicate items of income derived from sources other than regular activities
	(Example: interest earned)
line 22 —	Net income *(Loss)*
	(This amount is also transferred to item A, Schedule 2, Projected Cash Flow Statement)

Instructions - Projected Cash Flow, Schedule 2

This from is used to Project the flow of Cash by the Applicant for each year, from loan closing through the first full year of operation. Use the same number of columns as used on the Operating Budget, Schedule 1. These Cash Flow Projections are important in determining the adequacy of cash to cover operating expenses, transfers to debt payment, reserve accounts, etc.

Cash Basis Accounting

Applicants who maintain their records strictly on the cash basis of accounting and have no Accounts Receivable and Accounts Payable, may only need to complete the following line items: A, B-1, C-1, E-1 and E-3, F and G.

Line Item Instructions:

line A —	Bring forward the income or loss as entered on line 22, Schedule 1.
line B —	Add back any depreciation or other non cash items included on Schedule 1, Operating Budget.
line C —	Complete items C-1 through C-6 as appropriate, for item changes which provide for increase in cash balances. NOTE: Do not include changes in cash Accounts in Current Assets of item C. Lines C-3 and C-4 will indicate the changes in Working Capital *(Current Assets and Current Liabilities, Exclusive of Cash.)*
line D —	Enter the Net Total of all A, B and C items.
line E —	complete items E-1 through E-6 as appropriate for items for which cash was expended.
line F —	Enter the Beginning Cash Balance(s) for the period.
line G —	The total of item D less E-6 plus F will be the Ending Cash Balance(s). The total will be reconciled by balances in the various accounts, i.e., construction, revenue, debt, etc.

FIGURE A2.29 Sample operating budget.

Form RD 442-2 (Rev. 9-97)

Position 3
UNITED STATES DEPARTMENT OF AGRICULTURE
STATEMENT OF BUDGET, INCOME AND EQUITY

FORM APPROVED
OMB NO. 0575-0015

Schedule 1

Name _____ Address _____

For the _____ Months Ended _____

(1) OPERATING INCOME	PRIOR YEAR Actual (2)	ANNUAL BUDGET BEG _____ END _____ (3)	CURRENT YEAR Actual Data		Actual YTD (Over) Under Budget Col. 3 – 5 = 6 (6)
			Current Quarter (4)	Year To Date (5)	
1. _____					
2. _____					
3. _____					
4. _____					
5. Miscellaneous					
6. Less: Allowances and Deductions					
7. Total Operating Income (Add lines 1 through 6)					
OPERATING EXPENSES					
8. _____					
9. _____					
10. _____					
11. _____					
12. _____					
13. _____					
14. _____					
15. Interest					
16. Depreciation					
17. Total Operating Expense (Add Lines 8 through 16)					
18. NET OPERATING INCOME (LOSS) (Line 7 less 17)					
NONOPERATING INCOME					
19. _____					
20. _____					
21. Total Nonoperating Income (Add 19 and 20)					
22. NET INCOME (LOSS) (Add lines 18 and 21)					
23. Equity Beginning of Period					
24. _____					
25. _____					
26. Equity End of Period (Add lines 22 through 25)					

Budget and Annual Report Approved by Governing Body

Quarterly Reports Certified Correct

_____   _____   _____   _____
Secretary   Date   Appropriate Official   Date

According to the Paperwork Reduction Act of 1995, no persons are required to respond to a collection of information unless it displays a valid OMB control number. The valid OMB control number for this information collection is 0575-0015. The time required to complete this information collection is estimated to average 2-1/2 hours per response, including the time for reviewing instructions, searching existing data sources, gathering and maintaining the data needed, and completing and reviewing the collection of information.

FIGURE A2.30 Sample statement of budget, income, and equity. (continued)

Appendix 2

Page 2

SUPPLEMENTAL DATA
The Following Data Should Be Supplied Where Applicable

1. **ALL BORROWERS**   Circle One
   a. Are deposited funds in institutions insured by the Federal Government?   Yes   No
   b. Are you exempt from Federal Income Tax?   Yes   No
   c. Are Local, State and Federal Taxes paid current?   Yes   No
   d. Is corporate status in good standing with State?   Yes   No
   e. List kinds and amounts of insurance and fidelity bond: Complete <u>Only</u> when submitting annual budget information:

Insurance Coverage and Policy Number	Insurance Company and Address	Amount of Coverage	Expiration Date of Policy
Property Insurance Policy # _____	_____	_____	_____
Liability Policy # _____	_____	_____	_____
Fidelity Policy # _____	_____	_____	_____

2. **RECREATION AND GRAZING ASSOCIATION BORROWERS ONLY**   Current Quarter   Year to Date
   a. Number of Members   _____   _____

3. **WATER AND/OR SEWER UTILITY BORROWERS ONLY**
   a. Water purchased or produced (CU FT - GAL)   _____   _____
   b. Water sold (CU FT - GAL)   _____   _____
   c. Treated waste (CU FT - GAL)   _____   _____
   d. Number of users - water   _____   _____
   e. Number of users - sewer   _____   _____

4. **OTHER UTILITIES**
   a. Number of users   _____   _____
   b. Product purchased   _____   _____
   c. Product sold   _____   _____

5. **HEALTH CARE BORROWERS ONLY**
   a. Number of beds   _____   _____
   b. Patient days of care   _____   _____
   c. Percentage of occupancy   _____ %   _____ %
   d. Number of outpatient visits   _____   _____

6. **DISTRIBUTION OF ALL CASH AND INVESTMENTS***
   Indicate balances in the following accounts:

	Construction	Revenue	Debt Service	Operation & Maintenance	Reserve	All Others	Grand Total
Cash	$ _____	$ _____	$ _____	$ _____	$ _____	$ _____	$ _____
Savings and Investments	$ _____	$ _____	$ _____	$ _____	$ _____	$ _____	$ _____
Total	$ _____	$ _____	$ _____	$ _____	$ _____	$ _____	$ _____

7. <u>**AGE ACCOUNTS RECEIVABLE AS FOLLOWS**</u>:

	Days				
	0–30	31–60	61–90	91 and Older	*Total
Dollar Values	$ _____	$ _____	$ _____	$ _____	$ _____
Number of Accounts	_____	_____	_____	_____	_____

*Totals must agree with those on Balance Sheet.

FIGURE A2.30 Sample statement of budget, income, and equity. (continued)

PROJECTED CASH FLOW   Schedule 2

For the Year BEG. _____ END. _____
(same as schedule 1 column 3)

A. Line 22 from Schedule 1, Column 3 NET INCOME (LOSS) .... $ _____
    Add
B. <u>Items in Operations not Requiring Cash:</u>
    1. Depreciation (line 16 schedule 1) .... _____
    2. Others: _____ .... _____
C. <u>Cash Provided From:</u>
    1. Proceeds from Agency loan/grant .... _____
    2. Proceeds from others .... _____
    3. Increase (Decrease) in Accounts Payable, Accruals and other Current Liabilities .... _____
    4. Decrease (Increase) in Accounts Receivable, Inventories and
        Other Current Assets (<u>Exclude cash</u>) .... _____
    5. Other: _____ .... _____
    6. _____ .... _____
D. Total all A, B and C Items .... _____
E. <u>Less: Cash Expended for:</u>
    1. All Construction, Equipment and New Capital Items (loan & grant funds) .... _____
    2. Replacement and Additions to Existing Property, Plant and Equipment .... _____
    3. Principal Payment Agency Loan .... _____
    4. Principal Payment Other Loans .... _____
    5. Other: _____ .... _____
    6. Total E 1 through 5 .... _____
    Add
F. Beginning Cash Balances .... _____
G. Ending Cash Balances (Total of D Minus E 6 Plus F) .... $ _____

<u>Item G Cash Balances Composed of:</u>
    Construction Account .... $ _____
    Revenue Account .... _____
    Debt Payment Account .... _____
    O&M Account .... _____
    Reserve Account .... _____
    Funded Depreciation Account .... _____
    Others: _____ .... _____
    _____ .... _____

Total - Agrees with Item G .... $ _____

FIGURE A2.30 Sample statement of budget, income, and equity. (continued)

Schedule 1

STATEMENT OF BUDGET, INCOME AND EQUITY
INSTRUCTIONS

Community Program Borrowers

Frequency and Preparation:

1. When used as Management Report.
   (a) Prior to the beginning of each fiscal year, complete only column three, "Annual Budget," for the next fiscal year on page 1 and forward two copies to the County Supervisor. All data should be entered on the same basis as your accounting records, i.e., cash, accrual, etc. The budget must be approved by the governing body. Schedule 2, Projected Cash Flow will also be prepared and submitted at the same time.

   (b) Twenty (20) days after the end of each of the 1st 3 quarters of each year, complete all data on pages one and two and forward two copies to the County Supervisor. For 4th quarter Management Report, see (2) and (3) below.

2. When used as a year end Statement of Income. For borrowers not required to have an independent audit, and who are required to furnish Management Reports, complete all information on both pages of Schedule 1. This will serve as the 4th quarter Management Report and year end financial Statement of Income. This Annual Report will be approved by the governing body, with two copies submitted within 60 days of year end to the County Supervisor.

   For borrowers who are not required to furnish Management Reports, page 1 of schedule 1 may be used for the Annual Statement of Income if an annual audit is not required. In this case, complete only columns 1, 2 and 5.

Note: Year End Balance Sheet is also required in either of the aforementioned situations.

3. An independently audited Statement of Income containing budget and actual data will substitute for page 1 of this form as the 4th quarter Management Report, when required, and the year end Statement of Income. However, page 2 must be completed for all borrowers required to submit Management Reports.

Group Farmer Program Borrowers   1949-B (442.9)

Frequency and Preparation:

1. When used as Management Report submit Budget Data Only. Complete column three, "Annual Budget," for the next fiscal year on page 1, Schedule 1 and forward two copies to the County Supervisor. All data should be entered on the same basis as your accounting records, i.e., cash, accrual, etc. The budget must be approved by the governing body. When submitting along with Statement of Income, (item 2 below) include this budget data at the same time. Schedule 2, Projected Cash Flow is not required.

2. When used as year end Statement of Income. For borrowers not required to have independent audits, page 1 of Schedule 1 may be used for the Annual Statement of Income. Complete columns 1, 2 and 5. Also complete items 1, 6 and 7 on page 2. This form must be approved by the governing body, with two copies submitted within 60 days of year end to the County Supervisor. An independently audited Statement of Income will substitute for page 1, Schedule 1.

Column and Line Item Preparation

Column 1
Income and Expense Items:
All data entered should be on the same basis as your Accounting Record, i.e., Cash Basis, Accrual Basis, etc.

Operating Income
Lines 1 – 5   List types of income as appropriate.
Line 6   Allowances and Deductions
               (Pertains Generally to Health Care Institutions, and represents the difference between Gross Income and Amounts Received or to be Received from Patient and third party payors)

Operating Expenses
Lines 8 – 14   List types of expenses as appropriate
Line 15   Interest Agency
               (Interest expense incurred on Agency note(s).)
Line 16   Depreciation
               (Total depreciation expense for the year)
Line 18   Net Operating Income (Loss)
               (This amount represents the net operating income or loss before adding income not related to operations below)

FIGURE A2.30   Sample statement of budget, income, and equity. (continued)

INSTRUCTIONS - Column and Line Item Preparation Cont'd

Non Operating Income

Line 19 – 20 (Indicate items of income derived from sources other than regular activities, EX: interest, earned)

Line 22 Net Income (Loss)
(This amount is also transferred to item A of the Projected Cash Flow statement Schedule 2 when Management Reports are required for Community Program borrowers only.)

Line 23 Equity, Beginning of Period
(Enter the Equity at the beginning of Reporting Period. The term Equity is used interchangeably with Net Worth and Fund Balance.)

Lines 24 – 25 Enter items which cause changes in the Current Year's Equity other than line 22 amount.

Lines 26 Equity End of Period
(This balance will be the same amount that appears on the Balance Sheet.)

Column 2 - Prior Year Actual
Enter the actual income, expense and equity amounts of the prior year.

Community Program Borrowers: Use this column for all management report requirements except when submitting the proposed budget prior to the beginning of each fiscal year. Also fill in when using this Schedule as the year-end Statement of Income.

Group Farmer Programs: Complete only when also using this form as annual Statement of Income.

Column 3 - Annual Budget
This will be the budget for the current year when actual data is presented in columns four and/or five. When submitting only budget data on this form, the amounts will be for the next year. Enter the beginning and ending dates of the budget year at the top of this column.

Column 4 - Actual Data, Current Quarter
Only used by Community Program borrowers required to submit Management Reports and contains information for the current three months being reported.

Column 5 - Actual Data, Year to Date
For borrowers submitting Management Reports, enter cumulative data from the beginning of the Accounting Year through the Current Quarter. When used as Fourth Quarter Management Report and/or year end Statement of Income, enter data for the entire year.

Column 6 - Actual Year to Date (over) Under Budget
Only used by borrowers required to submit Management Reports and is determined by subtracting column 5 from column 3 for each line item.

SCHEDULE 1, PAGE 2, SUPPLEMENTAL DATA

This information is required of all borrowers submitting Management Reports. Fill in as indicated.

Community Program Borrowers complete as appropriate.

Group Farmer Program Borrowers complete only items 1, 6 and 7.

FIGURE A2.30 Sample statement of budget, income, and equity. (continued)

Schedule 2

PROJECTED CASH FLOW
INSTRUCTIONS

The completion of this form is required of all Community Program borrowers submitting Management Reports, and will accompany Schedule 1 when the Annual Budget is transmitted, to the County Supervisor. See Instruction No. 1 on Schedule 1.

This form is used to Project the Flow of cash for the budget year in order to determine the adequacy of cash to cover Operating Expenses, Transfer to Reserves, Debt Payment, Capital Outlays, etc.

Cash Basis Account - Systems
Borrowers who maintain their records strictly on the cash basis of accounting and have no Accounts Receivable and Accounts Payable, will probably only need to complete the following line items:
A, B-1, C-1, E-1 and E-3, F and G.

Line Item Instructions

Line A - Bring forward the income or loss as entered on line 22, schedule 1, column 3.

Line B - Add back any depreciation or other non cash items included on schedule 1, column 3.

Line C - Complete items C-1 through C-6 as appropriate, for item changes which provide for increase in cash balances.
Note: Do not include changes in Cash Accounts, in Current Assets of item C-4. Lines C-3 and C-4 will indicate the changes in Working Capital (Current Assets and Current Liabilities, Exclusive of Cash.)

Line D - Enter the net total of all A, B and C items.

Line E - Complete items E-1 through E-6 as appropriate for items for which cash was expended.

Line F - Enter the Beginning Cash Balance(s) for the Period.

Line G - The total of item D less E-6 plus F will be the Ending Cash Balance(s). This total will be reconciled by balances in the Various Accounts, i.e., Construction, Revenue, Debt, etc.

FIGURE A2.30 Sample statement of budget, income, and equity.

USDA
Form RD 400-8
(Rev. 8-00)

Position 5

FORM APPROVED
OMB No. 0575-0018

DATE OF REVIEW	COMPLIANCE REVIEW	STATE
		COUNTY
SOURCE OF FUNDS ☐ Direct ☐ Insured	(Nondiscrimination by Recipients of Financial Assistance through U. S. Department of Agriculture)	CASE NUMBER
		DATE LOAN OR GRANT CLOSED

TYPE OF ASSISTANCE
☐ Housing Preservation Grant   ☐ Water and Waste Disposal Loan or Grant   ☐ RRH and LH Organization
☐ RBEG   ☐ Grazing Association   ☐ Intermediary Relending Program
☐ RBOG   ☐ EO Cooperative   ☐ Rural Housing Site Loans
☐ B&I Loans   ☐ Community Facilities   ☐ Cooperative Service
                                                                            ☐ Other _____

NAME OF BORROWER ORGANIZATION OR ASSOCIATION

ADDRESS OF BORROWER

I. STATISTICAL INFORMATION

(For the purpose of this report, the term "PARTICIPANTS" will be used to describe "USER," "MEMBERS," OCCUPANTS," "SITE PURCHASER" OR Potential Users for pre-loan closing compliance reviews, as applicable.

A(I).

ETHNICITY	POPULATION		PARTICIPANTS			
			THIS REVIEW		LAST REVIEW	
	No.	%	No.	%	No.	%
Hispanic or Latino		0		0		0
Not Hispanic or Latino		0		0		0
TOTAL	0	100%	0	0	0	0
MALE		0		0		0
FEMALE		0		0		0

FIGURE A2.31   Sample compliance review. (continued)

A(2).

RACE	POPULATION		PARTICIPANTS			
			THIS REVIEW		LAST REVIEW	
	No.	%	No.	%	No.	%
American Indian/Alaskan Native		0		0		0
Asian		0		0		0
Black or African American		0		0		0
Native Hawaiian		0		0		0
White		0		0		0
TOTAL	0	100%	0	100%	0	100%
Male		0		0		0
Female		0		0		0

A(3).

EMPLOYEES

ETHNICITY			MALE		FEMALE	
	No.	%	No.	%	No.	%
Hispanic or Latino		NaN		NaN		NaN
Not Hispanic or Latino		NaN		NaN		NaN
TOTAL	0	NaN	0	NaN	0	NaN

BOARD OF DIRECTORS

ETHNICITY			MALE		FEMALE	
	No.	%	No.	%	No.	%
Hispanic or Latino		NaN		NaN		NaN
Not Hispanic or Latino		NaN		NaN		NaN
TOTAL	0	NaN	0	NaN	0	NaN

FIGURE A2.31 Sample compliance review. (continued)

A (3). cont. **EMPLOYEES** / **BOARD OF DIRECTORS**

RACE	No.	%	No. (M)	% (M)	No. (F)	% (F)
American Indian/Alaskan Native		0		0		0
Asian		0		0		0
Black or African American		0		0		0
Native Hawaiian		0		0		0
White		0		0		0
TOTAL	0	0	0	0	0	0%

RACE	No.	%	No. (M)	% (M)	No. (F)	% (F)
American Indian/Alaskan Native		0		0		0
Asian		0%		0		0
Black or African American		0		0		0
Native Hawaiian		0		0		0
White		0		0		0
TOTAL	0	0	0	0	0	0

II. APPLICATION INFORMATION (Project, Facility, Complex or Lender)

B(1).

ETHNICITY	Number of Application Received – This Review		Last Review		Number of Applications Approved		Number of Applications Rejected	
	No.	%	No.	%	No.	%	No.	%
Hispanic or Latino		0		0		0		0
Not Hispanic or Latino		0		0		0		0
TOTAL	0	0	0	0	0	0	0	0
TOTAL – Male		0		0		0		0
TOTAL – Female		0		0		0		0
TOTAL – Family		0		0		0		0

FIGURE A2.31 Sample compliance review. (continued)

B (1.)

RACE	Number of Application Received This Review		Last Review		Number of Applications Approved		Number of Applications Rejected	
	No	%	No.	%	No.	%	No.	%
American Indian/ Alaskan Native		0		0		0		0
Asian		0		0		0		0
Black or African American		0		0		0		0
Native Hawaiian		0		0		0		0
White		0		0		0		0
TOTAL	0	0	0	0	0	0	0	0
TOTAL Male		0		0		0		0
TOTAL Female		0		0		0		0
TOTAL Family		0		0		0		0

A. Are racial and gender of the participants and the number of employees in proportion to the population percentages? ☐ YES ☐ NO

B. Number of participants as of last review: _____ Date of last review: _____

C. Are all interested individuals permitted to file an application (written or otherwise) for participation? ☐ YES ☐ NO

If "NO" explain why not: _____

D. Does or will recipient of financial assistance maintain adequate records on the receipt and disposition of applications, including a list of applicants wishing to become participants? ☐ YES ☐ NO

If "NO" what action is being taken to establish adequate records: _____

If "YES" number of applicants wishing to become participants on list _____

Number on list from minority group _____

E. Number of applications received from prospective participants since last review: Total _____

If zero skip to III.

From minority group applicants _____

F. Number of applications which have been withdrawn or rejected since last review: Total _____
From minority group applicants _____

FIGURE A2.31 Sample compliance review. (continued)

G. Number of applications now pending on which no action has been taken: Total .... _____
   From minority group applicants.. .... _____

III. LOCATION OF THE FACILITY

A. Does the location of the facility or complex have the effect of denying access to any person on the basis of race, color, national origin, age, sex, or disability? .... ☐ YES   ☐ NO

B. Describe the racial makeup of the area surrounding the facility (if area is not the same as population).

IV. USE OF SERVICES AND FACILITIES

A. Are all participants required to pay the same fees, assessments, and charges per unit for the use of the facilities? .... ☐ Yes ☐ NO

   If "NO", explain:

B. Explain how charges for services, i.e., rent, connection, and user fees are assessed.

C. Is the use of the services or the facilities restricted in any manner because of race, color, or national origin? ....☐ Yes ☐ NO
   If "YES", explain:

D. Is there evidence that individuals, in a protected class, are provided different services, charged different or higher rate amounts than others? .... ☐ YES ☐ NO
   If "YES", explain:

E. List the methods used by the recipient to inform the community of the availability of services or benefits of the facility. (newspaper, radio, tv, etc.).

F. Do these methods reach the minority group population equally with the rest of the community? ....☐ Yes ☐ NO

G. Are appropriate Equal Opportunity posters conspicuously displayed? (And Justice For All and the Fair Housing poster)
   .... ☐ Yes ☐ NO

H. Do written materials, i.e., ads, pamphlets, brochures, handbooks and manuals, have a nondiscrimination statement, Fair Housing, and/or accessibility logo or Equal Opportunity statement? .... ☐ Yes ☐ NO

I. Describe the efforts of the recipient to attract minorities, females, and persons with disabilities to serve on the advisory board, board of directors, or similar boards.

J. Indicate whether the facility is being properly maintained and whether services are provided on a timely basis.

K. Describe any restrictions that may exist on the use of the facility, i.e., no playgrounds for children; restrictions on use by minorities, segregated or prohibited by age or disability of tenant or other participants.

FIGURE A2.31   Sample compliance review. (continued)

K. Describe any restrictions that may exist on the use of the facility, i.e., no playgrounds for children; restrictions on use by minorities, segregated or prohibited by age or disability of tenant or other participants.

L. If participation is restricted by age of beneficiary, please indicate any Federal statute, or state or local ordinance which may permit such restrictions.

M. How does this facility compare-with other similar facilities in the area serving low income beneficiaries which are privately or federally financed by other agencies.

Answer N for RRH and LH only:
N. Does the organization's Operating Rules provide for standard reasons for eviction? ☐ YES ☐ NO
If "YES," specify:

Are these reasons stipulated in the Lease Agreements? ☐ YES ☐ NO
If not, how are they made known to participants?

V. ACCESSIBILITY REQUIREMENTS (DISABILITY)
(For All Programs Funded By Rural Development)

A. Does the facility or project have an accessible route through common use areas? ☐ YES ☐ NO

B. Has a self-evaluation for Section 504 of the Rehabilitation Act been conducted and a transition plan developed for all structural barriers? ☐ YES ☐ NO

C. Does this facility or project have a Telecommunication Device for the Deaf (TDD) or participate in a relay service? ☐ YES ☐ NO

If not, is this part of the self-evaluation and transition plan? ☐ YES ☐ NO

D. Describe reasonable accommodations made by the recipient for making the program accessible to individuals with disabilities.

VI. ACCESSIBILITY REQUIREMENTS FOR RURAL RENTAL HOUSING

A. Does the complex meet the 5% accessibility requirement of 504 of the Rehabilitation Act of 1973 for facilities built after June 1982? ☐ YES ☐ NO

B. Are the units occupied by persons with disabilities in need of the special design features? ☐ YES ☐ NO

C. If not, indicate what outreach has been conducted utilizing appropriate organizations and advertising to reach the individuals in need of such units.

FIGURE A2.31   Sample compliance review. (continued)

VII. ACCESSIBILITY REQUIREMENTS FOR COMMUNITY FACILITIES
(Health Care Facilities)

A. List methods used by health care providers to communicate with the hearing impaired in the emergency room.

B. List methods used to communicate waivers and consent to treatment requirements to persons with disabilities, including those with impaired sensory or speaking skills.

C. Are there restrictions in delivery of services for the treatment of alcohol, drug addiction or other related illnesses?
(Aids, Hepatitis) ☐ YES ☐ NO

VIII. COMPLEXES AND FACILITIES THAT PROVIDE HOUSING
(Nursing Homes, Retirement Group, Rural Rental)

A. Does the facility have an approved Affirmative Fair Housing Marketing Plan? ☐ YES ☐ NO

B. Is there a copy of the most recently approved plan being used and conspicuously posted? ☐ YES ☐ NO

C. Is management meeting the objectives of the plan? ☐ YES ☐ NO

If not, is there an updated plan in place?

IX. PROGRAMS THAT CREATE EMPLOYMENT

A. Is there evidence that individuals in a protected class are required to meet different employment selection criteria than non-minorities? ☐ YES ☐ NO

B. Is there evidence that individuals of a protected class are being terminated in a disproportionate rate than non-minority employees? ☐ YES ☐ NO

C. Do recipients that employ fifteen or more persons have a designated person to coordinate its efforts to comply with Section 504 of the Rehabilitation Act of 1973? ☐ YES ☐ NO

D. Has the recipient provided reasonable accommodations to the known physical or mental impairment of employees with disabilities? ☐ YES ☐ NO

X. CONTACTS WITH INDIVIDUALS AFFILIATED WITH THE FACILITY OR COMPLEX

A. List contacts made with a diverse selection of tenants, users, patients, employees, and others affiliated with the facility or complex. List by name, race, sex, and disability (if provided).

B. Summarize comments made by the person(s) contacted.

FIGURE A2.31 Sample compliance review. (continued)

XI. COMMUNITY CONTACTS

A. List contacts made with community leaders and organizations representing minorities, females, families with children, and individuals with disabilities. Include the date and the method of contact.

B. Summarize comments made by person(s) contacted.

XII. PAST ASSISTANCE FROM RD OR OTHER FEDERAL AGENCY

A. List past loans or other federal financial assistance from other agencies.

B. Does the recipient have a pending application with RD or another Federal agency? ☐ YES ☐ NO

XIII. CIVIL RIGHTS COMPLIANCE HISTORY
Provide a history of the following:

A. Compliance Review. Has this recipient had a finding of non-compliance by RD or another federal agency? ☐ YES ☐ NO

B. Discrimination Complaints. Has a complaint of prohibited discrimination been filed against this recipient in the past three (3) years? ☐ YES ☐ NO

C. Law Suit. Has a law suit based on prohibited discrimination been filed against this recipient in the past three (3) years? If so, describe and attach copies of the law suit. ☐ YES ☐ NO

D. Did the recipient take appropriate corrective or remedial action to achieve compliance with civil laws or to resolve any discrimination complaint cases or law suits? ☐ YES ☐ NO

E. Identify the resources and or contacts used in verifying the recipient's past civil rights compliance history.

FIGURE A2.31 Sample compliance review. (continued)

XIV. CONCLUSIONS

A. Did your review of the records maintained by the association or organization disclose any evidence of discrimination on the grounds of race, color, national origin, sex, age, or disability in the services or use of the facility? .... ☐ Yes ☐ NO
   If "YES," describe in detail such discrimination:

B. Did your contacts with community leaders, including minority leaders, disclose any evidence of discrimination as to race, color, national origin, sex, age, or disability in the services or use of the facility? .... ☐ Yes ☐ NO

C. Did your observation of this borrower's operations or proposed operations indicate any discrimination on the grounds of race, color, national origin, sex, age, or disability in the services or use of the facility? .... ☐ Yes ☐ NO
   If "YES," describe in detail such discrimination:

D. Comments for other observations or conclusions:

Based upon my observation of this borrower's operation or proposed operation and the attitude of the Governing Body and Officials it is my opinion that the Recipient _____ Is _____ Is Not complying with the requirements under Title VI of the Civil Rights Act of 1964, Section 504 of the Rehabilitation Act of 1973, Age Discrimination Act of 1975, and Title IX of the Education Amendments Act of 1972.

_____   _____
DATE   COMPLIANCE REVIEW OFFICER

XV. RECIPIENT IS IN NON-COMPLIANCE (Complete only if there is a finding of non-compliance)

A. Sent recipient notice of non-compliance on this date _____

B. Date of compliance meeting _____

C. Target date for recipient to voluntarily comply _____

D. Recipient has complied with all requirements and made all necessary corrective action by this date _____

E. Describe all meetings with recipient to achieve compliance.

F. Recipient has refused to voluntarily comply by this date _____

G. Comments:

FIGURE A2.31 Sample compliance review.

Appendix 2

BOARD OF ZONING APPEALS
FOR ALLEGANY COUNTY, MARYLAND

CASE #
DATE FILED
HEARING DATE

APPEAL FOR VARIANCE UNDER ZONING ORDINANCE

Appeal is hereby made pursuant to §141-129.D.2. of the Code of Allegany County, Maryland, for a Variance from a requirement or requirements of that Ordinance, as follows:

Subject Property M-_____, Q-_____, P-_____ Lot(s)_____, Zoning Classification_____

Location Description_____

Applicant's present legal interest in the subject property: ( ) Owner
( ) Lessee
( ) Contract to ☐lease/☐rent/☐purchase

Ordinance section from which a variance is desired:_____

Nature and extent of variance:_____

Due to the lot's ( ) narrowness ( ) shallowness ( ) shape ( ) area ( ) topographical characteristics ( ) geology ( )soils ( ) proximity to floodplain ( ) _____, addressing the subject requirement(s) would result in a peculiar and unusual difficulty and cause exceptional and/or undo hardship to the Applicant.

Describe extraordinary situation or exceptional topographical conditions of the property:_____

If exceptional narrowness, shallowness or shape of property is claimed, give date plat recordation of present subdivision or state that a deed describing the lot was recorded prior to March 3, 1972._____

Has there been any other Appeal involving the subject property?:_____

NOTARY: STATE OF _____,
County of _____, TO WIT:

I hereby certify that on this ____ day of _____, 200____, before me the subscriber, a Notary Public of the State aforesaid, personally appeared _____, known to me, or satisfactorily proven to be the person(s) whose name is subscribed to the foregoing instrument, who acknowledged that he has executed it for the purposes therein set forth, and that it is his act and deed.
In witness thereof, I have set my hand and Notarial Seal, the day and year first written above.

_____
Notary Public
My Commission expires on _____ 200____

I hereby agree to comply with all regulations and codes, which are applicable hereto. I further agree that any misstatement or misrepresentation of facts presented as part of this application, or change to proposal without approval of the agencies concerned, shall constitute sufficient grounds for the disapproval or revocation of the subject permit. I hereby affirm that I own the property which is the subject of this application; or that I am the duly designated representative of the property owner, and that I possess the legal authority to make this Affidavit on behalf of myself or the owner for whom I am acting. I do solemnly declare and affirm under the penalties of perjury that the contents of this Application are true and correct to the best of my knowledge, information and belief.

Applicants Signature_____
Address_____
_____
City/State/Zip_____

\\boa\forms\BOA Petition for Variance.doc
rev 9/02 [E-3110]

FIGURE A2.32 Sample appeal for zoning variance form.

**PLACER COUNTY
PLANNING DEPARTMENT**
11414 "B" AVENUE, AUBURN, CA 95603
(530) 889-7470

PLANNING APPEALS

The specific regulations regarding appeal procedures may be found in the Placer County Code, Chapters 19 (Subdivision), 30 (Planning and Zoning), and 31 (Environmental Review Ordinance).

--Office Use Only--

Last Day to Appeal _____ (5 p.m.)
Letter _____
Oral Testimony _____
Zoning _____

Appeal Fee $_____
Date Appeal Filed _____
Receipt # _____
Received by _____
Geographic Area _____

--To Be Completed by Applicant--

1. Project Name: _____
2. Appellant(s): _____
3. Address: _____ Phone: _____
4. Assessor's Parcel Number(s): _____
5. Application Being Appealed (check all those that apply):
   ___ Development Agreement #DAG - _____
   ___ Use Permit #CUP- _____
   ___ Parcel Map #P- _____
   ___ General Plan Amendment #GPA- _____
   ___ Specific Plan #SPA- _____
   ___ Interpretation of
         Planning Director: _____ (date)
   ___ Other: _____
   ___ Tentative Map #SUB- _____
   ___ Variance #VAA- _____
   ___ Design Review #DSA- _____
   ___ Rezoning #REA- _____
   ___ Rafting Permit #RPA- _____
   ___ Env. Review EIAQ- _____
   ___ Minor Boundary
         Line Adjustment #MBR- _____

6. Whose Decision is Being Appealed: _____
   (see reverse)
7. Appeal to be Heard By: _____
   (see reverse)
8. Reason for Appeal (attach additional sheet if necessary and be specific):
   _____
   _____
   (If you are appealing a project condition only, please state the condition number.)

Note: Applicants may be required to submit additional project plans/maps.

Signature of Appellant(s) _____
_____

(attach additional sheet for signatures if necessary)

FIGURE A2.33 Sample planning appeal form. (continued)

PLACER COUNTY ZONING ORDINANCE
SECTION 25.140

APPEALS TO THE PLANNING COMMISSION, THEN TO THE BOARD OF SUPERVISORS

Rulings made by:

- Planning Director (interpretations)
- Zoning Administrator (ZA)
- Design/Site Review Committee (D/SRC)
- Parcel Review Committee (PRC) (other than road improvements which should be appealed to the Director of Public Works)
- Environmental Review Committee (ERC)

APPEALS DIRECTLY TO THE BOARD OF SUPERVISORS

Rulings made by:

- Planning Commission

APPEALS TO THE HEARING BODY HAVING ORIGINAL JURISDICTION

Rulings made by:

- Development Review Committee

NOTE: AN APPEAL MUST BE FILED WITHIN TEN CALENDAR DAYS OF THE DECISION. APPEALS FILED MORE THAN TEN DAYS AFTER THE DECISION SHALL NOT BE ACCEPTED BY THE PLANNING DEPARTMENT. (Section 25.140 C.1., Placer County Zoning Ordinance)

T:\CMD\CMDP\LORI\APPS\APPEAL
7/96

FIGURE A2.33 Sample planning appeal form.

Your Company Name
Your Company Address
Your Company Phone and Fax Numbers

INDEPENDENT CONTRACTOR ACKNOWLEDGMENT

Undersigned hereby enters into a certain arrangement or affiliation with Your Company Name, as of this date. The Undersigned confirms:

1. Undersigned is an independent contractor and is not an employee, agent, partner or joint venturer of or with the Company.

2. Undersigned shall not be entitled to participate in any vacation, medical or other fringe benefit or retirement program of the Company and shall not make claim of entitlement to any such employee program or benefit.

3. Undersigned shall be solely responsible for the payment of withholding taxes, FICA and other such tax deductions on any earnings or payments made, and the Company shall withhold no such payroll tax deductions from any payments due. The Undersigned agrees to indemnify and reimburse the Company from any claim or assessment by any taxing authority arising from this paragraph.

4. Undersigned and Company acknowledge that the Undersigned shall not be subject to the provisions of any personnel policy or rules and regulations applicable to employees, as the Undersigned shall fulfill his/her responsibility independent of and without supervisory control by the Company.

Signed under seal this _____ day of _____, 20 ___.

_____   _____
Independent Contractor   Company Representative

                                                                           _____
                                                                           Title

FIGURE A2.34 Sample independent contractor acknowledgment form.

Appendix 2

CONTRACTOR RATING SHEET

Job name: _____ Date: _____

Category	Contractor 1	Contractor 2	Contractor 3
Contractor name			
Returns calls			
Licensed			
Insured			
Bonded			
References			
Price			
Experience			
Years in business			
Work quality			
Availability			
Deposit required			
Detailed quote			
Personality			
Punctual			
Gut reaction			

Notes: _____

_____

_____

_____

_____

_____

FIGURE A2.35 Sample contractor rating sheet.

SUBCONTRACTOR LIST

Service	Vendor	Phone	Date

FIGURE A2.36 Sample contractor list.

Appendix 2

Your Company Name
Your Company Address
Your Company Phone and Fax Numbers.

LONG-FORM LIEN WAIVER

Customer name: _____

Customer address: _____

Customer city/state/zip: _____

Customer phone number: _____

Job location: _____

Date: _____

Type of work: _____

The vendor acknowledges receipt of all payments stated below. These payments are in compliance with the written contract between the vendor and the customer. The vendor hereby states that payment for all work done to this date has been paid in full.

The vendor releases and relinquishes any and all rights available to said vendor to place a mechanic or materialman lien against the subject property for the described work. Both parties agree that all work performed to date has been paid for, in full and in compliance with their written contract.

The undersigned vendor releases the customer and the customer's property from any liability for nonpayment of material or services extended through this date. The undersigned contractor has read this entire agreement and understands the agreement.

Vendor*	Services	Date Paid	Amount Paid
Plumbing Contractor	Rough-in		
Plumbing Contractor	Final		
Electrician	Rough-in		
Electrician	Final		
Supplier	Framing Lumber		

*This list should include all contractors and suppliers. All vendors are listed on the same lien waiver, and sign above their trade name for each service rendered, at the time of payment.

FIGURE A2.37 Sample long-form lien waiver.

Your Company Name
Your Company Address
Your Company Phone and Fax Numbers

EARLY TERMINATION AND MUTUAL RELEASE OF CONTRACT

For good and valuable consideration had and received and the mutual promises and releases herein contained, the parties known as _____ (Contractor) and _____ (Customer) do hereby release each other, now and forever, in and from all further promises, liabilities, warranties, requirements, obligations, payments, and performance of the contract dated _____, 20 _____, entitled _____ and made for the purpose of _____ _____ as reflected in said contract between them.

    The parties each acknowledge all matters between them regarding the said contract have been satisfactorily adjusted between them, and the contract has been terminated prior to its entire fulfillment and performance, as the parties have agreed such early termination is mutually desirable.

    Accordingly, said contract is hereby SUPERSEDED AND ABSOLUTELY TERMINATED.

    Each party warrants each's own full power and authority to enter into this Early Termination and Mutual Release of Contract, which shall become effective only upon the signature of both parties.

Date: _____ Date: _____

Customer: _____ Contractor: _____

by: _____(Seal) Title: _____

                                                    by: _____(Seal)

State of _____ of _____

The foregoing Early Termination and Mutual Release of Contract was sworn to and acknowledged before me by _____ and _____ on _____, 20 _____.

_____
Notary Public

My commission expires:_____   (Notary Seal)

FIGURE A2.38 Sample early termination and mutual release contract.

Your Company Name
Your Company Address
Your Company Phone and Fax Numbers

CODE VIOLATION NOTIFICATION

Contractor: _____

Contractor's address: _____

Contractor's city/state/zip: _____

Contractor's phone number: _____

Job location: _____

Date: _____

Type of work: _____

Subcontractor: _____

Address: _____

OFFICIAL NOTIFICATION OF CODE VIOLATIONS

On _____, 20 ____, I was notified by the local code enforcement officer of code violations in the work performed by your company. The violations must be corrected within _____ (___) business days, as per our contract dated _____, 20 ____. Please contact the codes officer for a detailed explanation of the violations and required corrections. If the violations are not corrected within the allotted time, you may be penalized, as per our contract, for your actions in delaying the completion of this project. Thank you for your prompt attention to this matter.

_____   _____
Developer   Date

FIGURE A2.39 Sample code violation notification.

Company Name
Your Company Address
Your Company Phone and Fax Numbers

WEEKLY EXPENSE REPORT

Employee name: _____ Department: _____
Office location/extension: _____ Week ending: _____

TRANSPORTATION	Sun.	Mon.	Tues.	Wed.	Thurs.	Fri.	Sat.	TOTAL
Total automobile miles x mileage rate .20								
Gas, oil, maintenance								
Parking and tolls								
Auto rental								
Taxi								
Other (air, rail, bus)								
TOTAL TRANSPORTATION								
MEALS AND LODGING								
Hotel (include parking, tips)								
Breakfast								
Lunch								
Dinner								
Other Meals								
TOTAL MEALS AND LODGING								
MISCELLANEOUS								
Laundry, cleaning								
Phone, fax								
Sundries								
Entertainment (detail below)								
TOTAL MISCELLANEOUS								
PER DAY TOTALS								

Total expenses _____
Less cash advance and charges to Company _____
Amount due me (Company) $ _____

Entertainment Details

Date	Event	Clients Entertained	Location	Business Purpose	Amount

Signature: _____ Title: _____ Date: _____
Approval: _____ Title: _____ Date: _____

FIGURE A2.40 Sample weekly expense report.

Your Company Name
Your Company Address
Your Company Phone and Fax Numbers

INDEPENDENT CONTRACTOR AGREEMENT

I understand that as an Independent Contractor I am solely responsible for my health, actions, taxes, insurance, transportation, and any other responsibilities that may be involved with the work I will be doing as an Independent Contractor.

I will not hold anyone else responsible for any claims or liabilities that may arise from this work or from any cause related to this work. I waive any rights I have or may have to hold anyone liable for any reason as a result of this work.

_____   _____
Independent Contractor   Date

_____   _____
Witness   Date

FIGURE A2.41 Sample independent contractor agreement form.

Subdivison Regulations to Consider

Topic	Considered	Acceptable	Unacceptable	Need More Data
Zoning				
Easements				
Deed restrictions				
Covenants				
Road access				
Public sewer				
Public water main				
Public gas main				
Tap fees for connecting to public utilities				
Street design				
Walkways				
Signage				
Green space requirements				
Erosion protection				
Environmental regulations				
Building restrictions				
Minimum lot-size requirements				

FIGURE A2.42 Subdivision regulations to consider.

Components of Construction Plans for Land Development

Component	Have	Need	Don't Need	Ordered
Cover sheet				
Site conditions				
Vicinity map				
Symbol legends				
Developer's name				
Development name				
Grading plan				
Site plan				
Existing building plans				
Proposed building plan				
Road plans				
Utility plans				
Parking areas				
Storm water drawings				
Existing contours				
Proposed contours				
Sediment and erosion plan				
Landscaping plan				
Easement details				

FIGURE A2.43 Components of construction plans for land development.

Components of Final Project Development Plans

Component	Included	Not Yet Included	Not Needed
Project name			
Project section (for large developments)			
Surveys			
Legal description of property			
Names of property owners			
Right-of-ways			
Easements			
Names of streets			
Lot addresses			
Covenants			
Restrictions			
Building plans			
Restricted areas			

FIGURE A2.44 Components of final project development plans.

Appendix 2

Agency Approvals That May Be Needed

Agency	Need	Don't Need	Have
Zoning department			
Environmental Protection Agency			
Department of Environmental Protection			
Soil conservation district			
Public works department			
Department of Education			
Fire and rescue department			
Building codes enforcement			
Parks and recreation department			
Local specific utility agencies			

FIGURE A2.45 Agency approvals that may be needed.

Components for a Project Design

Component	Need	Don't Need	Need More Data	Have
Project drawings				
Site plan				
Elevation plan				
Surveys				
Overview plan				
Zoning maps				
Topographical maps				
Transportation drawings				
Utility plans and maps				
Road designs				
Proposed lot sizes and location				
Landscaping plan				
Housing plan				
Recreational area plans				
Greenspace plans				
Common area plans				
Waterways and pond plans				

FIGURE A2.46 Components for a project design.

Components of a Preliminary Plan

Component	Checked	Approved	Not Approved	Not Applicable
Soils test				
Geology reports				
Zoning requirements				
Site plan				
Defined land use				
Street design				
Traffic signage				
Utility plan				
Easement verifications				
Wastewater plan				
Lot sizes and locations				

FIGURE A2.47 Components of a preliminary plan.

Types of Permits That May Be Required

Permit Required	Not Required	I Already Have It
Clearing		
Demolition		
Road entry		
Land use		
Dredging		
Environmental		
Private septic system		
Private water system		
Street cutting		
Building		
Plumbing		
Heating		
Electrical		
Occupancy		

FIGURE A2.48 Types of permits that may be required.

Appendix 2

Water Considerations			
Consideration	Applies	Does Not Apply	Acceptable Risk
Pollution			
Flood risk			
Waterways in development area			
Erosion			
Public water availability			
Private well potential			
Public sewer availability			
Private sewage disposal			
Retainage water potential			
Stormwater			

FIGURE A2.49 Water considerations.

Types of Environmental Risks to Consider

Risk	Applies	Does Not Apply	Acceptable	Unacceptable
Hazardous waste				
Toxic substances				
Asbestos				
Radon				
Underground storage tanks				
Pesticides				
Wetlands				
General pollutants				

FIGURE A2.50 Types of environmental risks to consider.

Environmental Considerations

Consideration	Applies	Does Not Apply	Acceptable Risk
Soil studies			
Water studies			
Pollution studies			
Air quality			
Noise			
Health risks			
Adjoining and adjacent properties			
Traffic studies			
Wildlife studies (any endangered species?)			

FIGURE A2.51 Environmental considerations.

Characteristics to be Wary of					
Characteristic	Exists	Does Not Exit	Is Ruled Out	Is a Risk Factor	Acceptable Risk
Stream					
River					
Pond					
Lake					
Ferns					
Cattails					
Wet areas					
Swales					
Erosion					
Culverts					
Retaining walls					
Ridges					
Bedrock					
Potential flood zone or plain					
Visible refuse					
Signs of hazardous dumping					
Vandalism					
Condition of adjacent and adjoining property					
Road noise					
Traffic congestion					
Apparent existing utilities					

FIGURE A2.52 Characteristics to be wary of.

Appendix 2

Potential Types of People Needed for a Project

Type of Person	May Need	Definitely Need	Already Have	Don't Need
Real estate attorney				
Lender				
Real estate appraiser				
Insurance company				
Bonding company				
Surveyor				
Engineering firm				
Landscape architect				
Project designer				
Environmental specialists				
Heavy-equipment company for clearing land				
Tree-cutting experts				
Road construction contractor				
Structural architect				
Building contractor				
Sales staff or agent				
Marketing expert				
Advertising agency				

FIGURE A2.53 Potential types of people needed for a project.

Considerations for Evaluating a Project

Consideration	Applies	Does Not Apply	Is Satisfactory	Is Not Satisfactory
Time needed for project development				
Location of project				
Demographics				
Zoning				
Setbacks				
Deed restrictions				
Wetlands				
Erosion potential				
Hazardous materials				
Soils studies				
Comparable properties				
Schools				
Shopping				
Public transportation				
Local housing				
Employment opportunties				
Traffic count				
Recreational opportunities				
Code requirements				
Is the site in a historic district?				
Comparable sales				
Present real estate market conditions				
Projected real estate market conditions				

FIGURE A2.54 Considerations for evaluating a project.

Index

Accountants, 21, 74
Appraisal process, 165–166
Architects, 22
Attorneys, 21
Auctions, 72

Back taxes, 72
Backup plans, 184
Bids, 143–146
Brokerage sales, 29–30
Brownfield redevelopment, 78
Budgets, 176–178
Builders, 21–22, 40–41, 199–201, 202–203
Building plans and specifications, 135–136
Building regulations, 64–65

Carpenters, 201–202
Changes, 136
Civil engineers, 22–23
Closed drainage systems, 83–84
Closing process, 164–165
Cluster housing, 159–160
Commercial banks, 37–38
Communication, 182
Compaction equipment, 93–94
Comparing bids, 144–145
Competition, 6–7

Complex roads, 100–102
Construction plan, 135
Construction regulations, 65–67
Contractors, 40–41 139–144
Control, 5, 183
Covenants , 194–195
Creative terms, 41–43
Credit reports, 167–168
Cumulative zoning, 157–158
Cutting costs, 137

Dealing directly, 70–72
Demographics, 124–126
Demolition plan, 133–134
Development property, 45–58
    Appraisers, 52–53
    Availability, 46–47
    Average deals, 55
    Competition, 54–55
    Defining your needs, 47–48
    Do-it-yourself, 52
    Fine tuning, 57–58
    Going around in circles, 58
    Hard sells, 54
    Mass market, 56–57
    Setting a budget, 49–51
    Working with brokers, 51

Development team, 19–25, 196
Drainage, 81–87
    Closed systems, 83–84
    Looking at the land, 86
    Open systems, 84–86
    Sectional designs, 86–87

Environmental concerns, 118–119
Environmentally Friendly Developments, 187–197
    Considerations, 188–189
    Covenants and restrictions, 194–195
    General design, 190
    Grass, 191
    Mixed-use ideas, 191–192
    Partners, 189
    Pedestrians, 190
    Preservation, 192
    Ride-sharing parking lots, 191
    Street solutions, 190–191
    Wastewater treatment, 193–194
    Wind breaks, 193
Erosion factors, 79
Estimates, 140–142
Exclusionary zoning, 160

Failed projects, 128
Features, 10–11
Fire hydrants, 109–110
Floating zones, 158
Flood areas, 117
Foreclosures, 73

General design, 190
Grade plan, 134
Grass, 191
Green space, 102–103
Growth status, 128–129

Hazardous waste, 118
Hype, 152–153

Inclusionary zoning, 160–161
Income verifications, 167
Independent sales, 28
Investors, 35–41
Invoices, 177

Irrigation, 79

Key considerations, 188–189

Land, 69
    Accountants, 74
    Auctions, 72
    Back taxes, 72
    Dealing directly, 70–72
    Foreclosures, 73
    Lawyers, 74
Land and improvements, 151–152
Land loss, 97–103
    Complex roads, 100–102
    Green space, 102–103
    Natural road sites, 100
    Simple access roads, 98–100
Land only, 150–151
Land use, 63–64
Lawyers, 74
Legwork, 182–183
Lenders, 25–26, 35–41
Listing agreements, 30–31
Local trends, 148–148
Locating contractors, 142–143

Marketing, 32–33
Marketing consultants, 21
Master plan, 60–63
    Components, 60
    Small developments, 61
    Your job, 61–63
Mixed-use properties, 78, 191–192
Money, 2, 10, 11–12, 15–16, 183–184
Mortgage brokers, 38–39

Natural road sites, 100

Open drainage systems, 84–86
Options, 16
Organizational skills, 16–17

Partners, 4–5, 41, 189
Pedestrians, 190
People skills, 18
Plan layers, 131–133

Planned growth, 78
Planned unit developments, 159
Preservation, 192
Price increases, 178
Private investors, 39–40
Project manager, 67
Project planners, 22
Proximity research, 127–128

Real estate team, 149–150, 195
Red tape, 7–8
Refining bids, 145–146
Reports, 177
Research, 17
Restrictions, 194–195
Ride-sharing parking lots, 191
Risk, 6
Routing water mains, 108–109

Sales, 27–32, 31–32, 147–148
Sales projections, 147–148
Scheduling, 178–185
Sectional designs, 86–87
Seller financing, 15–16
Seller participation, 16
Simple access roads, 98–100
Site needs, 66
Site supervision, 171–174
Small developments, 61
Soil, 89–95
    Assessment, 95
    Bearing capacity, 91–92
    Compaction equipment, 93–94
    Stability, 94–95
    Types, 90–91
Soliciting bids, 143–144

Space preservation, 77–78
Street solutions, 190–191
Supervision, 180–182
Surveyors, 23
Surveys, 167

Tapping into water mains, 110–111
Temporary construction, 65–66
Time management, 18–19
Title search, 166
Tracking, 168–169
Traffic plan, 134–135
Traffic studies, 126–127
Transitional zoning, 158

Urban infill, 78
Utility plan, 134

Wastewater treatment, 193–194
Water demand, 106–107
Water mains, 108–109
Wells, 111–112
Wetlands, 115–117
Wind breaks, 193

Zoning, 155–161
    Cluster housing, 159–160
    Cumulative, 157–158
    Exclusionary, 160
    Floating, 158
    Inclusionary, 160–161
    Other types, 161
    Planned unit developments, 159
    Transitional, 158
Zoning maps, 155–157